DATE DUE

Writing the Future

Terra Nova Books aim to show how scientific issues have cultural and artistic components. Combining essays, reportage, fiction, art, and poetry, Terra Nova Books reveal the complex and paradoxical ways the natural and the human are continually redefining each other.

Other Terra Nova books:

Writing on Air

Writing on Water

The New Earth Reader

The World and the Wild

The Book of Music and Nature

Terra Nova
New Jersey Institute of Technology
Newark, NJ 07102
973 642 4673
terranova@njit.edu
www.terranovabooks.org

Writing the Future

Progress and Evolution

edited by David Rothenberg and Wandee J. Pryor

A Terra Nova Book

The MIT Press
Cambridge, Massachusetts
London, England

This book was set in Berkeley Old Style Book by Graphic Composition, Inc., using QuarkXPress, and was printed and bound in the United States of America.

Library of Congress Cataloging-in-Publication Data

Writing the future : progress and evolution / edited by David Rothenberg and Wandee J. Pryor.
p. cm.
"A Terra nova book."
Includes bibliographical references.
ISBN 0-262-18235-1 (alk. paper)
1. Science—Literary collections. 2. Science. I. Rothenberg, David, 1962– II. Pryor, Wandee J.

PN6071.S3W75 2004
808.8′036—dc22

2003066624

Printed on recycled paper.

10 9 8 7 6 5 4 3 2 1

Contents

II *Steps from the Cave*

III *Places in Time*

Where Are We Going, Where Have We Been?

David Rothenberg and Wandee J. Pryor

Evolution is the single clearest theory that connects humanity to the natural world and explains how and why we are a part of it. It is brilliant in its simplicity, ubiquitous in its reach. Before it, nature is wild, multifarious, and confusing, and afterward, everything might still be a mess, but there is a reason for the mess—an explanation for the circuitous paths that got life where it is today. Adaptive behavior, the purposeful result of accident. It is John Cage's view of art writ large upon the world: nature's complexity is remarkably the result of chance operations. The peacock's unwieldy tail is as useful as the ant's body design is efficient. All qualities have been selected because they are the most successful—even things that seem exaggerated and strange.

Once you start to think about it, the range of natural selection spreads outward, and many imagine it can explain everything. Nature evolving, now no mystery. Every wonder of life has its purpose, based on simple desires: survival, competition, sex, proliferation. The intricate machine-like couplings of moths, the cruelty of chimpanzees, the strange preferences of weird female and male beasts alike. Does selection actually explain this? And, where, in this era of cloning, genetic manipulation, and biotechnical enhancement, do we humans fit in? Can the theory of millions of years of living evolution have any use when applied to the narrow time frame of human civilization, the motion in time we like to call progress?

Evolution by natural selection is the one scientific revolution of the past two hundred years that can be grasped at the human scale, as it describes the human place in nature and immediately changes the way we see the world. That is why it is a fitting subject for a *Terra Nova* book. As such a compelling and comprehensible vision, it has been as easily abused as it has been invoked to explain all sorts of nuances of human behavior, from war to business, art to exploitation. In this collection, we examine both practical and speculative sources of biological and social phenomena. Through these far-ranging essays, poems, and images, we see how the concepts of evolution and progress can be useful in appreciating our own shifting place in the world.

Each of the four parts of this anthology gathers pieces around a distinct aspect of our overall theme. In part I, "Moths, Sex, and Chaos," we include creative investigations of some of the founding concepts and personalities of evolution and natural selection, focusing on the scientific theories of adaptation and evolutionary biology.

Theodore Roszak begins the book with a challenge to Darwinian orthodoxy, which he traces back to the other pioneer of the theory, Alfred Russel Wallace. Roszak questions evolution's inability to explain how the human mind became so sophisticated. Despite all our civilization's advances, we still don't do a good job explaining what the brain can do, and the metaphors of machines offered by human progress aren't much help either. We think the mind is evolution's finest achievement, while the rest of the planet stands by mute as we admire ourselves.

Michael Ruse questions the progress we assume must be happening in the unfolding of evolutionary theory. He believes that evolution's history is no more progressive than is evolution itself, presenting a notion of science as highly biased to the cultural values that shift over the decades. Competition? Cooperation? What you see in nature depends on what you look for. The key to grasping the real value of science, he surmises, is admitting that it is a creation of culture and society.

In his essay on sex, David Geary explains that evolution is not really about progress, but about adapting, across many generations, to ecological and social change. The same could be said about our own explanations for our and nature's behavior. We circle around mysteries that are never transcended by anything we say. We wish to be creatures of free will, ever deciding what to do next, what to create, and what to destroy. Now if all our individual decisions are just blips in some erratic slow motion of the species that moves with no goal in sight, we too get amazing results without needing to know where we're supposed to end up.

Part II, "Steps from the Cave," focuses on this adaptability, the changes in humankind. It includes pieces on personal, internal human change, as well as the inevitable experiences of maturing and aging. It swings from tales of destruction and growth to assessing the ways humanity has evolved in language and in art. It considers how evolution and progress play out as metaphors and methods to make sense of our own, specific lives. Ellen Dissanayake takes us into the utter strangeness of the French caves that contain the earliest paintings, going back at least 30,000 years. What touches her most is the alien quality of these spaces, which she imagines would have been just as strange way back then. Only human beings, she surmises, could have taken pleasure in such surreal, uncomfortable environments, using them to express their greatest stories.

Floyd Skloot's story of Isaac is a richly layered tale of body, mind, family, and dreams. What takes Isaac's focus from basketball to God? Only a sudden journey to death and back to life. Science may explain his condition but cannot say what to do with this new-found knowledge. Isaac looks elsewhere and sees a way. Valerie Hurley must come to terms with the tragic death of her child and the massively destructive impulses in her own culture. Nature and society change, surprising us, going nowhere in particular. We must sift through our own experiences to grasp glimmers of meaning and hope.

We move into part III, "Places in Time," looking for answers outside ourselves, delving into the details of our surroundings that morph as change takes place. No longer focused on internal experience, we look at the world evolving. The focus is now outside. Leslie Van Gelder moves from memories of her father, the curator of the American Museum of Natural History, through a series of remarkable places, seeking to articulate how place is relived and reexperienced through time and the tales one tells.

Eva Salzman turns a mundane walk through a Long Island neighborhood into a sonata of ancient language, myth, and modern memory. David Petersen eloquently defends what some would call an isolated mountain life, while he links to the world through the happenings of nature and the call of the wild. Carolynne Baker digs deep into the layers of history in a Vietnamese city and tries to decode human experience. We have gotten here from there, choosing pictures and explanations so the places we go can fit into us, in both the past and the future.

Now progress, what's that? An idea a bit older than evolution but not so much. The sense that human civilization is on the rise, improving, going somewhere rather than nowhere. That as we live, we improve our lives, with things getting better all the time.

This is a much more limited, though paradoxically harder, thing to prove than evolution by natural selection. Why? Because it involves the notions of better and worse, and defining those matters of aesthetics has always been tough at best, and certainly unscientific. When it comes to creatures of nature, no one has ever said we are better than the dinosaurs. We may have large brains and be able to remake our world in any image we like, but the great beasts are certainly impressive and genuinely sublime. Science cannot say *Homo sapiens* is preferable to a lyrebird or slime mold, or a butterfly better than a moth.

Is progress more or less controversial than evolution? Its veracity doesn't divide true believers from the holders of cold, hard facts. We are supposed to live longer than our grandparents, and we will drive cars that are more plastic and automated. They never touched a computer, and everything we call work happens through those wry digital beasts. We can link with the far corners of the world in an instant, but they spoke eight languages, not us. We are at home nowhere, if everywhere, but they better understood what it meant to belong.

There are good old days and bad old days. The new decays faster than the ancient, so a thousand years from now, what will be left? And is there enough wildness left to let evolution continue, heading off somewhere to some grand future where human beings will not matter much, if even remain? No species is supposed to be on top forever. Soon we will run out of some essential resources our ancestors had absolutely no need for, as we whirl on to the end of the industrial world, hoping to reach the beginning of something else: a way of life kinder, fairer, wiser, more careful, and more aware. But is that where we're headed? Is that the goal progress has set itself up to reach?

In part IV, "Getting to the Future," we consider the concept of progress and wonder where that fickle piper might lead, what wild trajectories our species and our planet might travel together in the years to come. Is progress in human civilization ever unambiguously manifest? Here are distinct visions of how the present is moving and pictures on what is found to be progress. Kathleen Creed Page suggests that new reproductive technologies radically transform what it means to invent the next generation. Yet as we tame the life force, we must recognize its autonomy and never get carried away with our own powers.

Kevin Warwick literally put a chip on his shoulder: he had a small electronic circuit implanted under his skin so that his own nervous system could be linked to a computer. He is quite optimistic about where this will get us. With brains literally plugged into computers, there will be no need to work so hard or learn so much. Whatever we need to know can be automatically downloaded into us. Joan Maloof agrees that technology might bring us to such cyborg human-machine hybrids, but wonders instead if the real future for our species is a kind of deevolution, where we slowly return to nature and abdicate our dominating role. Once we have seen the possibilities, we may in the end choose humility. Andy Couturier reports on one Mr. Nakamura, a Japanese wood carver who studied with a Tibetan master and lives a simple life in his home constantly carving, keeping an ancient tradition going. He is part of the modern world, but not pushing it onward to something that it is not. He chooses simplicity and adapts only within himself and his art.

As human beings, it is much easier for us to talk about progress than about evolution, because we clearly know why our future is indefinite while for the science of nature, we're supposed to trust the experts. Evolution, they tell us, is a theory both intuitive and counterintuitive. "Don't you see?" they say. "Natural selection allows life to be cryptic and common at once, producing beautiful variety without any guiding hand being in charge." Make all human behavior a subset of biology, the way E. O. Wilson consiliently wants it to be. Or prove that very simple algorithms can lead to the most astonishing of patterns, as Stephen Wolfram and other chaoplexity theorists have, reforming science on the simplest step-by-step mathematical grounds.

But we are not here to talk only about science. We are interested in how the ideas of evolution and of progress stir artists into action, challenging all to place their work in the largest scheme of palpable change, asking, "Where do I fit into the march of time?" Evolution slowly unveils the deep mysteries deep time has wrought upon living forms, and progress is the dream that humanity can make this motion go on, continually improving the direction it takes. The older you get, the more you realize things do not stay the same. You realize that society's collective span of attention is short. What seemed known by all is quickly forgotten, and the same mistakes are made. Progress? All we know is that change doesn't hold still.

There is more information today than any of us could master. There are more years to live and fewer reasons to live for. There are faster computers and slower decisions, and the rapid decimation of species that have taken millennia to finesse. The only reason for so much human-made change is the hope that progress will solve, inspire, transform us and the world into some better place, moving all the while with no one in charge.

With no guide at the helm, progress looks suspiciously like evolution in nature, even though, in biotopic time, all of our problems are a blip like a fly in the Taj

Mahal. This might offer a glimmer of hope: if nature can supply wonderfully elegant solutions to the problems of survival by trying out test models derived solely by chance, then surely it's possible for us to find our way forward. Evolution, bolstered by accident, moving in fits and starts, does not rush things. You won't see it whizzing by at the speed a computer might be able to simulate. Patience is the tempo out there, in the world beyond us. Nature renews itself, moving in cycles, but is going nowhere, with nowhere to go.

We all speak of change. We make decisions and plans, concoct schedules and time lines. We steer our life in specific directions even when, in the back of our minds, there's a gnawing sense that the future stays beyond our control. The experts tell us not to be fooled, not to sit so smug, that no one ever said human beings were the cream of the crop, the end product of eons of randomness and accidents finessed. Selection produced us, and all the beauty of nature besides, but nothing had us in mind when it began.

We can accept that; indeed, it gives us a lot less to worry about. But we're still ill at ease when it comes to progress. Ever since the word got out and education reached beyond the royal and the religious, with the great burst of the Enlightenment, we have read, thought, and changed the way we live in search for a better life. That's the story we've all been told for years. Why is it, then, that the same question, "What is the good life?" hangs over us all the way down from the ancient Greeks? We are no closer than they were to figuring out how to frame an answer, even though we pat ourselves on the back and call ourselves more enlightened, more respectful, and open to those who are different. Philosophy does not progress, nor does literature or art. There are glimmers of greatness amid acres of waste, and we ought always to train ourselves to look for what will last. Yet beauty is fleeting, cleverness fleeting, timeliness fleeting.

Our minds and bodies are equally restless. "So you wanna direct evolution?" laughs biologist Lynn Margulis. "Just have as many children as you possibly can." So is procreation the only evolutionary influence a person can have? Ideas appear, fan out, get caught, make waves or fade, touch some and mostly go unnoticed. We're puffed up like bullfrogs if we think that we actually matter to the world. Who else watches TV? Who marks down the march of time? The rest of nature is too sure of itself for such doubt.

Science must by its own definition be careful and scrupulous, never loose with its claims. As much as we talk of unlocking the mysteries of origin, how the past could turn into the present, the way experiments lead to accepted knowledge is always a slow and meticulous game. Mysteries are to be savored, danced around, evoked with the indefinable influence of image and word.

Evolution shows the great power of aimless change. But human survival, if we make it, is something to celebrate only after we take aim. Not to fire, not to hit, but to envision and guess the new way out. We like that line, but you know, this *terra nova,* this new world, has always been there, even before our search began. We just didn't know how to get there, but maybe we are wise enough not to sail blindly into the unknown. We are blessed with the adaptive need to ask questions, and the wisdom not to be satisfied with easy answers. Art answers the hardest questions best, because its solutions don't make the questions go away, but keeps them enduring forever in that part of ourselves that wants to hold on to the great, the impossible, and the true.

I

Moths, Sex, and Chaos

Wallace's Dilemma: Evolution and Transcendence

Theodore Roszak

Where is evolution going?

For most biologists, the answer is: everywhere in general and nowhere in particular. Standard Darwinian theory understands evolution to be a free-for-all governed by random combinations of DNA and lucky environmental selection. Living things evolve in all directions, opportunistically occupying any niche for which their genetic roulette prepares them. "Survival of the fittest" means the survival of those that "fit" the shifting ecological contours of their habitat. Within the terms of that paradigm, there is room for serious debate. Darwin held that the process of selection and adaptation was gradual but, given enough time, capable of remarkable feats of transformation. He was prepared to believe that a bear might have been reshaped into a whale. Others, like Richard Goldschmidt, the American geneticist, have suggested that the sudden appearance of "hopeful monsters" may account for radical new departures in evolution—the big jumps that give rise to new genera. Still others, like Stephen Jay Gould, have insisted that catastrophic events and "great dyings" play a central and largely randomizing role in accelerating the emergence of new species. But all orthodox biologists agree that the process operates nonteleologically, meaning without any purpose in view. Evolution is ceaseless, shapeless, aimless.

Darwin had no trouble believing the "struggle for existence" has no greater goal than brute survival. The cultural climate of his time made it easy. As every textbook reminds us, he created his theory of natural selection during the high noon of dog-eat-dog capitalism. Thomas Malthus and the free market economists were never far from his thoughts. At one point, he observed that everything he knew about the breeding of plants and animals fell into place as a theory of evolution when he read Malthus's "Essay on the Principle of Population." Ironically, Malthus, who insisted (by way of a precise mathematical formulation, no less) that population must outrun the food supply, has been proved wrong by history. Similarly, classical economics, so inspiring to Darwin, has long since come under heavy critical fire. But evolutionary biology remains linked to the image of ordered randomness once seen in the free market. Natural selection is the biological version of Adam Smith's "invisible hand." Both ideas lay claim to being "scientific" because they purport to be value free. But that does not mean they are philosophically neutral. Classical economics was invented to chase government regulation from the marketplace; similarly, Darwinism was quickly seized on by militant atheists to drive God from the universe. Nature, the Darwinians argued, runs by itself; no central planning was needed.

Biological laissez-faire continues to dominate mainstream science, but there have always been significant doubts. Alfred Russel Wallace, cofounder of natural selection, was among the earliest dissenters. He agreed that natural selection explained adaptation, but in his eyes, adaptation was essentially conservative and unenterprising. It moved in a purely horizontal direction, molding plants and animals to their environment in ever more specialized, and so inflexible, ways. In his 1889 book, *Darwinism,* Wallace raised an issue he and Darwin were to debate for years. Wallace's key point of reference was the human brain. The brain broke the rules. It was, so Wallace believed, "an instrument [that] has been developed in advance of the needs of its possessor." Wallace seized on a key aspect of evolutionary theory: "Natural selection could only have endowed the savage with a brain a little superior to that of the ape, whereas he actually possesses one very little inferior to that of the average member of our learned societies." These days, some would argue that primary people, whose cultures have endured longer than our own, may have more wit, more horse sense, more wisdom than their civilized conquerors. In any case, Wallace's dilemma is both simple and devastating. If evolution unfolds as a series of graduated modifications, each passed along either because it is neutral in effect or competitively advantageous, how are we to account for an organ like the brain that so vastly exceeds what might be required to outsmart any existing primate rival?

For Wallace, the mind overarched natural selection. He believed there was a more daring, vertical movement that boosts life toward higher levels of complexity and consciousness. After all, if evolution were merely a matter of survival by adaptation, we might still be a planet of hearty bacteria. Those bacteria would have their history, an eons-long series of variations and adaptations, all responsive to selection, but without movement toward greater complexity. For that matter, if complexity beyond the unicellular level were rare and episodic, coming and going over the eons, we would still have evolution as Darwin explained it. But in the only example we have of evolving life—our own Earth—we see something more dramatic. We see a steady, undeterred thrust toward a net gain in complexity. The microbes continue, but life has branched out into an amazing array of new species. It has been building itself up into ever more delicate, sentient forms. To ignore that fact would be to ignore the defining feature of evolution. Suppose we had before us a scrambled selection of pictures, each showing a life form ranging from the microbe to the dolphin. We would know that these pictures belonged in a certain obvious order. We would know they described a history in which the simpler came first and the more complex followed. If it could be shown that this were not the case—if, for example, we had reason to believe the elephant appeared before the lungfish, or the whale before the crab—think how that would undermine our understanding of evolution.

If one leaves out the thrust toward complexity as a necessary element of evolution, how does one account for the emergence of sexual reproduction? Sex is a riskier form of reproduction than employed by the microbes. Bacteria have done quite well for billions of years without dividing into male and female. Those who believe "selfish genes" are the main story of evolution might do well to wonder why anything exists beyond microbes, the most gene-retentive form of life. By way of sex, each living thing gambles away half its genes, sharing the remainder in unpredictable combinations. Why should sex have endured and prospered, unless the unpredictability and complexity it makes possible were somehow part of the evolutionary story?

Issues of complexity grow even more pressing when we reach the level of mentality. As Wallace asked, what is the evolutionary status of art and music? What is the status of science? "The special features we have been discussing clearly point to the existence in man of something which he has not derived from his animal progenitors—something which we may best refer to as being of a spiritual essence or nature, capable of progressive development under favorable conditions. . . . Thus alone we can understand the constancy of the martyr, the unselfishness of the philanthropist, the devotion of the patriot, the enthusiasm of the artist, and the resolute and persevering search of the scientific worker after nature's secrets."[1]

Do these uniquely human qualities perhaps point toward a destined goal beyond physical survival and reproductive advantage? Fascinated by the transcendent impulse of the mind, Wallace clung to the conviction that "a superior intelligence has guided the development of man in a definite direction and for a special purpose, just as man guides the development of many animal and vegetable forms."[2] In his later years, Wallace was drawn to spiritualism and parapsychology as possible keys to human nature, interests that eventually cost him his credibility among biologists. Perhaps that is why contemporary biology has given Wallace's dilemma only minor attention. By and large, Darwinians fail to see any validity in his great question. As with the body, so with the mind. Steven Pinker (*How the Mind Works*) and Daniel Dennett (*Darwin's Dangerous Idea*) speak for mainstream evolutionary theory when they insist that the mind was built up incrementally by way of small, selective advantages in the same way as a bird's wing.[3] They see the growth of intelligence as wholly a matter of problem solving and toolmaking—practical talents to which natural selection easily applies. They simply ignore Wallace's dilemma, offering no reason why the mind should ever have developed beyond simple counting, toolmaking, and enough verbal ability to coordinate a hunting expedition. Today the assumption seems to be that once technological intelligence gets started, there will be no stopping it. It simply takes over the mind's development. Darwin, in his *Descent of Man,* had a similar notion. He replied to Wallace that the mind of primitive man had to grow larger to accommodate such apparently inevitable capacities as the use of language, tools, fire, and the wheel.

Not surprisingly, those who subscribe to such a technological imperative have a very clear idea of where evolution is heading. As early as the mid-1970s, Carl Sagan was convinced that machine intelligence, consciousness, and free will were "inevitable." Thus, *The Dragons of Eden,* Sagan's study of the evolution of human intelligence, finishes with an enthusiastic chapter on artificial intelligence. "The next major structural development in human intelligence," Sagan believed, "is likely to be a partnership between intelligent humans and intelligent machines."[4] Similar enthusiastic predictions that cyborgian alliances and machine intelligence are the way forward in evolution are now commonplace in popular science and predictions of the future. Ideas like this produce marvelously eye-popping computer graphics on the screen. Conceive of the mind as a data processor, and computers are bound to look like rather promising mechanical minds—possibly better than the human original. A computer calculates faster, files more data, follows logical rules more accurately. It uses words and numbers with unambiguous precision; it does not sleep, dream, lie, forget, goof off, or go crazy. Is it not everything a really good mind should be?

There are many computer scientists who would agree. "Our minds are our brains," says Dennett, "and hence are ultimately just stupendously complex 'machines.' The difference between us and other animals is one of huge degree, not metaphysical kind."[5] Dennett does not prove this; he simply asserts it. What else can he do? There is no way to prove the validity of a metaphor. "The moon is a ghostly galleon," says the poet. "The mind is a machine," says the biologist. The metaphor comes easily to those who start by choosing the computer as a model of intelligence. This is the line of thought that led cognitive scientists to believe that a chess-playing computer would prove that the mind could be mechanically mimicked, if not surpassed. I suspect that those who started out along this track never once doubted that they had the right model. Did any of them ask if mothering children was a better example of intelligence than winning at chess?

We have since learned that human beings don't play chess the way computers do; and even if they did, the logic of chess pales beside the complexity of what we call "common sense," the capacity we exercise 99 percent of every day when we go shopping, brush our teeth, watch a ball game, argue a political issue, or raise our children. That may be the most valuable lesson that thinking machines have taught us about thinking: namely, that we don't think like machines. Not that we can't be trained to imitate computers. Tell children that they ought to think like computers, and you will produce a race of human beings who are indeed inferior to the machines they use.

Our growing awareness of the difference between thinking machines and thinking minds has not discouraged some experts in artificial intelligence from seeing computers as the wave of the future. There seems to be no hesitation about bringing values and goals into the evolutionary story if one does so to serve a technological end. There are those who believe the best data processors to come may not be made of flesh and blood. It is not only science fiction that now flirts with the possibilities of human obsolescence. Imagine, at today's rate of progress, two or three more centuries of research in artificial intelligence and genetic self-replication. Imagine the two fields of study coalescing into one science. What wonders of transhuman evolution might then be within our technological reach!

In the view of MIT's Marvin Minsky, one of the most aggressively reductionistic experts in artificial intelligence, "The amount of intelligence we humans have is arbitrary. It's just the amount we have at this point in evolution. There are people who think that evolution has stopped, and that there can never be anything smarter than us."[6] Minsky has called the brain a "meat machine," which, like all machines, can be analyzed, adjusted, and improved upon. In the same vein, Robert Jastrow of NASA believes that "human evolution is nearly a finished chapter in the history of life. . . . That does not mean the evolution of intelligence has ended on the earth. . . . We can expect that a new species will arise out of man, surpassing his achievements as he has surpassed those of his predecessor homo erectus. The new kind of intelligent life is more likely to be made of silicon."[7] Jastrow thinks this evolutionary leap to sentient computers may still be a million years off—a safe prediction. Ray Kurzweil, author of *The Age of Spiritual Machines,* believes our human obsolescence is closer at hand. He predicts "the emergence in the early twenty-first century of a new form of intelligence on Earth that can compete with and ultimately significantly exceed human intelligence."[8] Before the century ends, we may have surrendered to our digital rivals, transferring our minds into computers that will grant us a sort of electronic immortality. As I recall from my Catholic boyhood, the reward of life everlasting was a gift only God could bestow.

Predictions like those of Jastrow and Kurzweil reveal a kind of high-tech Manichaeanism that has long been an underlying theme of modern science: the hope of liberating pure reason from the physical facts of life—and incidentally from the messy bodily intimacies of sex. Note the assumption, as Jastrow puts it: "Mind is the essence of being."[9] Delete the body, and identity remains intact. At the birth of modern Western philosophy, Pythagoras and Plato seized upon mathematics as the purest expression of deathless being. Two thousand years later, at the beginning of the modern era, Descartes echoed that same desire to rise above the flesh when he separated calculating mind from corruptible matter and made mathematics the official language of science. As an essentially computational machine, the computer has inherited the flight from mortality as a subliminal goal that continues to cast its spell over many of the brightest minds in the world of high tech. Here we have the Cartesian dictum, "Cogito, ergo sum," pressed to its literal and logical extreme. "I" become nothing other or more than my cogitating brain. If, therefore, that brain can be simulated in silicon, "I" survive. Mark Slouka has this same strange alliance of the ascetic and the mathematical in view when he characterizes high tech as "an attack on reality as human beings have always know it." Cyberspace, he believes, is getting crowded with evolutionary scenarios about uploading consciousness into electronic networks. Behind these high-tech fantasies he sees "a fear and loathing of the natural world, of physical experience in its entirety."[10] The astrophysicist Frank Tipler has pressed these possibilities even further. In his book *The Physics of Immortality: Modern Cosmology and the Resurrection of the Dead,* he posits the evolutionary equivalent of the Christian resurrection. "The dead," he tells us, "will be resurrected when the computer capacity of the universe is so large that the amount of capacity required to store all possible human speculations is an insignificant fraction of the entire capacity." Humanity would then dominate the entire universe; we would have progressed "from Earth-womb into the cosmos at large."[11]

Long before that far horizon is reached, Tipler is certain that we will be able to simulate the body in all its most refined details—and improve upon it. It would then be unnecessary to preserve the carnal original; it might be cast aside in favor of its robotic equivalent. Such an "emulated person," Tipler argues, "would observe herself to be as real, and as having a body as solid as the body we currently observe ourselves to have."[12] The simulated body would, however, have one very special quality: it would be deathless. In that form, disembodied minds of the future might be loaded aboard a spacecraft and fired off into the universe to explore the galaxies far, far away—needing no air, food, water, or exercise for the journey. Even boredom need not be a problem; a disincarnate intelligence need be placed in a comatose state only for the thousands of years it may take to arrive at a destination light-years away.

In the pages of *Wired* magazine, silicon immortality is among the constant themes of the cyberpunk intelligentsia. This may in fact be the emotional subtext for the advanced claims of artificial intelligence. Interviewed in *Wired,* Chris Langton, one of the founders of artificial life research, puts it this way: "There are these other forms of life that want to come into existence. And they are using me as a vehicle for reproduction and for implementation."[13] Vernor Vinge, looking further into the future, tells us, "If we ever succeed in making machines as smart as humans, then it's only a small leap to imagine that we would soon thereafter make—or cause to be made machines that are even smarter than any human. And that's it. That's the end of the human race within the animal kingdom."[14]

Not all computer scientists endorse these predictions. Bill Joy of Sun Microsystems sees the prospect nearer at hand than many others do and dreads it. Contemplating a future that "doesn't need us," Joy believes that by 2020, we will be trapped in the grip of autonomous robots that have learned how to remodel themselves and will be using human beings as they see fit—a vision straight out of Karel Capek's play *RUR*. "A second dream of robotics is that we will gradually replace ourselves with our robotic technology, achieving a near immortality by downloading our consciousness. . . . But . . . what are the chances that we will thereafter be ourselves or even human?"[15] Jaron Lanier, creator of virtual reality and a maverick member of the computer community, sadly concludes that biorobotic fantasies and visions of "spiritual machines" are actually among the major attractions of computer science. He agrees that many hackers "nurture hopes of being able to live forever by backing themselves on to a computer tape." He characterizes these ambitions as the beginning of a new "zombie culture" dominated by ex-humans who "are ready to leave all that behind and imagine living on a disk in which they only interact with other minds and environmental elements that also exist solely as software." This, Lanier believes, is what accounts for that curious new psychological category we call "nerdiness."[16] Intellectually, the nerd is one who searches for ways to digitalize away all distinctions of quality, feeling, and affect. Emotionally, the nerd is given over to an alien blandness that wants shelter from human intimacy and physicality.

There are other, less nihilistic views of the mind in mainstream evolutionary biology. Stephen Jay Gould sides with the linguistic philosopher Noam Chomsky. For Chomsky and Gould, the essence of the mind is language; linguistic capacity is what seems to have built the brain so big and made it so cortically convoluted. Neither Gould nor Chomsky believes language can be accounted for by selective advantage. Chomsky holds that the "language organ" or the "language acquisition device (LAD)" is innate in a way that eludes biological explanation. That's why every child acquires language with an almost instinctive rapidity. Language, Chomsky suggests, is embedded in the deep, mental capacity to generate sentences. It assumes the form of a universal grammar that has not yet been mapped in the brain—or, for that matter, fully elaborated in theory. Could such a grammar have evolved incrementally in the same way that upright posture evolved? Chomsky doubts it, and his doubt is well taken. No other creature has managed to elaborate its repertory of communicative sounds into a fully developed grammar. There are no half-articulate beings, though Darwin, in one of his less brilliant moments, believed there might once have been. He speculated that there were once quasi-talkers, creatures possessed of a "half-art and half-instinct of language." He reasoned that "the continued use of language will have reacted upon the brain and produced an inherited effect: and this again will have reacted on the improvement of language. The large size of the brain in man . . . may be attributed in chief part . . . to the early use of some simple form of language."[17]

Leaving aside the strikingly Lamarckian implications of this hypothesis—that the exercise of speech genetically transformed the brain—it should be possible to support this odd notion with an example of such a "simple form of language." True, there is animal communication, but warning grunts and courting songs—"instinctive cries," as Darwin put it—do not make sentences—which leaves us to wonder how Chomsky's LAD could come into existence if not all at once—perhaps as a kind of quantum phenomenon. Just as there are limited numbers of orbits an electron can occupy inside the atom, perhaps there is no resting place for intelligence between the chimp level and the human level. And just as we have no idea how an electron vanishes from one orbit and reappears in another, so we may never know how mentality made the jump from preverbal to verbal. Such a quantum leap of the mind is difficult to imagine, but no more so than trying to imagine the gradual, piecemeal evolution of grammar. Can we envision a time when there were verbs without nouns, adjectives without adverbs, subjects without predicates? And if that is not what we mean by a "simple form of language," what *do* we mean?

Pondering the mystery of human speech, Stephen Jay Gould wonders if language, and with it the whole of culture, might be the spin-off of some other ability chosen by natural selection. "Yes," he says, "the brain got big by natural selection. But as a result of this size, and the neural density and connectivity thus imparted, human brains could perform an immense range of functions quite unrelated to the original reasons for increase in bulk."[18]

It is truly remarkable the lengths to which even brilliant scientists will go to in order to defend the prevailing orthodoxy. Let us take Gould's proposition seriously as the basis of a thought experiment. If we cannot imagine language evolving into existence directly by virtue of selective advantage, can we imagine it developing in all its richness as a mere by-product of something else? Well, what practical ability can we imagine appearing in the preverbal "struggle for existence" that might have simply lapped over into the hieroglyphs of Egypt or the Homeric epics? Mating calls perhaps? Love serenades have continued to fulfill their purpose in numerous species without developing beyond the variations we hear in melodic birdsong. By now, we might have expected some clever canary to have spun off the basis for a subordinate clause or a future tense, but we find no obvious connection with sentence structure among the world's loveliest singers. Building a fire or whittling a spear has survival value. But along what lines can we imagine fire or spear making being transformed into sonnets and scientific theories? How? Why?

If we can speculate so freely, why not try something more ambitious? Might there have been some primordial form of highly articulated communication that served as the template for spoken language and then vanished, leaving behind a pattern that was taken over by speech as we know it? Geneticists have wondered if many of the oddities of DNA arise from the possibility that DNA took over from some more primitive replicator and carried forward characteristics that made chemical sense in that earlier version. Perhaps the same is true of spoken language. I can think of one other human capacity that has a complex grammar of sorts, and that is dancing. Not something most linguists know much about. Yet anybody who has watched a dance teacher or choreographer at work clearly sees that the anatomy of the body—which is, after all, universal—dictates a precise grammar of movement. There are steps that must grow out of others and movements that cannot be linked together. There are ways the body can bend and stretch, and ways in which it cannot. Dance results from a collection of inarticulate rules that arise from the way muscles, bones, and joints work under the influence of gravity and within the limits of human strength and balance. The vocabulary of steps and the grammar of movement might qualify as a "simple form of language." Once dance had developed into a flexible form of communication and expression—perhaps before the tongue and larynx had been adapted to articulation—it might have come to be paired with vocal "synonyms," which at last replaced the choreographed version. The original sentence structure may have been dance phrases linked together in ways that were communicative, efficient, and graceful. Perhaps Chomsky's transformational grammar is so elusive because it was once something else: a logic of movement. This is, of course, little more than a guess. But then all theories about the origins of language are guesswork. At least with this line of thought, we are reminded that communication, including the communication of complex ideas, transcends the spoken word and that the aesthetics of movement may once have counted for as much as the practicalities of physical survival.

The best that standard biology can do with the brain's unaccountable excursion into cultural creativity is to give it a name: *hypertrophy*. Excess—perhaps the sort of excess that often proves fatal, as may have been the case with the dinosaurs, which went to extremes in body weight. But a name is not an explanation. And there is surely something odd about so dismissive a treatment of the very mind that brought forth scientific inquiry. If we value the quest for truth, as every scientist must, are we to regard the brain that searches for truth as no more than a luxurious surplus of electrochemical circuitry that has spent most of its history dealing in dreams and fantasies, mathematical puzzles and metaphysics? Of what practical value is high-energy physics or cosmology? What contribution do these marvelous fields of speculation and research make to brute survival? And what of evolutionary theory itself? Does evolutionary theory, one of our most delightful brainchildren, make some generous place for the mind that may now shape the future course of evolution? That is surely so startling a possibility that it deserves to be assessed as more than accidental excess.

The Robinson Crusoe–Tom Swift image of mind that most biologists and cognitive scientists hold dear is good, solid eighteenth-century science. John Locke, David Hume, and Benjamin Franklin would have heartily approved. This is the mind of *homo faber,* ideally bereft of dreams and unsettling visions, never in need of psychiatry or spiritual counseling. Yet on strict Darwinian principles, one could make a case that the real mind we share with our ancestors across the eons is as much a liability as a blessing. Think of the delusions and neuroses it has brought with it, the nightmares and manias. For every practical skill intelligence has given us, there are countless fantasies, myths, delusions, folktales, and superstitions that seem to make no contribution to our physical survival. Why then did rational intelligence ever give rise to—or preserve—the unconscious mind? In the evolution of efficient intelligence, why would the burden of neurosis not have long since been selected out? Recall that both Freud and Jung traced the roots of culture to the unconscious mind, and both believed we learn more about our nature from the symbolic transformations of myth, art, and religion than from strict logic. William James wondered if great religious truths might not enter the mind by the "trap door" of the unconscious. No serious student of the psyche has regarded the unconscious as excess biological baggage. Rather, it has been seen as the essence of human personality, the mystery at the core of our being. The dream defines us at least as much as the digging stick does.

If the unconscious and the irrational are somehow connected to a useful function, it cannot be something as simple as practical problem solving or information processing, but something as strange and elusive as the unfolding cosmos we inhabit, a place of bewildering emergent possibilities. Following Wallace, countless evolutionary philosophies have pondered the place of mind in nature. All have agreed that it is the frontier of evolution, which is, admittedly, a self-serving view. The whales and the oak trees are in no position to dispute the role we assign ourselves as the vanguard of life on Earth. We announce that status, but only the silence of our fellow species surrounds us. Yet the claim need not be made arrogantly, nor need it ignore the hazards and responsibilities that befall pioneers. It can, indeed, be a humbling and civilizing lesson to see ourselves at the forefront of a grand, cosmic vista that dwarfs the selfish passions and petty distractions of the moment and may be unfolding toward possibilities far beyond anything we have yet envisioned.

But it is one thing to decide that the mind is the leading edge of evolution, another to decide what *mind* most essentially means. Whose mind do we choose as our model? Here is where the controversy deepens. Scientists understandably cast human nature in their own intellectual image, preferring the analytical and empirical habits that characterize their professional life. Are we justified in taking language to be the key to mentality? Is the mind most basically an organ of communication? It is tempting for scientists and academics to see things that way. These are highly articulate people, and the skill they most value—logic—is grounded in the rules of linguistic argument.

Outside the small, busy world of cognitive science, philosophers of evolution have celebrated many other dimensions of mind. Nietzsche and George Bernard Shaw envisaged the evolutionary superman as artist and philosopher. Teilhard de Chardin believed it is the saints who will usher us to the culmination of human development. The systems theorist Erich Jantsch (in *Design for Evolution*) regards love and the "feminine element" as the rejuvenating force of human evolution.[19] Henri Bergson placed mystic intuition at the forward edge of the *Èlan vital.* He argued that the task of the mystic (whom he saw as an emergent new species) is to humanize technology so that it might liberate us from material necessity for a higher, religious calling. "Man will rise above earthly things only if a powerful equipment supplies him with the requisite fulcrum. He must use matter as a support if he wants to get away from matter. In other words, the mystical summons up the mechanical."[20]

To emphasize, as these philosophers do, the evolutionary role of the compassionate, the creative, the mystical is a useful corrective to the current fascination with computerized intelligence. It reminds us that where the means of mass destruction have reached so awesome a level, our survival may depend more on the saints who set humane goals than on the technicians who provide ingenious means. Norbert Wiener, the founding father of cybernetics, knew as much; he warned us that "know what" comes before "know how."[21]

The discussion of mind is complicated by the fact that many who claim to be experts in the field—as if they alone were aware of their thinking brains—are part of a centuries-long debunking campaign that has become a matter of scientific orthodoxy. They are in the tiresome habit of being tough and reductive about everything, as if their very manhood depended on it. No soft, sentimental ideas for them. "Nothing but . . . nothing but . . . nothing but." But even such one-upmanship eventually has to face the paradox that haunts every discussion of the mind. In mathematics, Gödel's theorem of incompleteness states that the axioms of any formal system cannot be wholly proved from within the system itself. Thus, no logical system can ever come full circle and bite its own tail. Cognitive scientists differ in their evaluation of Gödel's theorem. Some, like Daniel Dennett, think it has no place in the study of intelligence. But the theorem can be generalized beyond mathematics. In a larger cultural sense, it might be called the paradox of transcendent context. Think something, and at once there is more that lies outside that something and serves as a frame. Any idea we hold in mind has to be smaller than the mind that holds it, for the mind encompasses it, contemplates it, compares it with other ideas, criticizes it, perhaps finds it wanting and decides to cast it away. Our history is a scrap heap of outmoded ideas that have come to be regarded as false, foolish, or inadequate. Some of what we have discarded were once somebody's Theory of Everything. But the minds of others got around those theories and found more attractive possibilities.

Among the ideas that fill our cultural repertory are ideas about the mind itself, and that fact introduces a startling convolution into our understanding. Evolution is a process we say accounts for the mind that formulated evolution. The biologist's notion that the mind is a product of evolution arises within the mind itself. There is a mind that is thinking this idea of the mind. And that mind transcends the mind it holds as an idea. The same is true of the cognitive scientist's notion that the mind is a mushy sort of computer. How, then, can that mind be placed inside a machine that the mind has invented and decided to use as a model of itself?

The mind is bigger than logic and mathematics—bigger than any machine it invents. But it is just as important to realize that the mind is bigger than art and religion as well. It is bigger than anything we can stand away from and view critically as an option—which is, quite simply, every element of human culture. Indeed, the mind is so big that we cannot see its boundary any more than we can see the ever-receding edge of the universe. If there is any name we can assign to the boundary of mind, it has nothing to do with information, calculation, logic, or knowledge. It would have to do with that realm of life we call "experience," that spontaneous reaching out of all our senses and capacities toward something *more*—more wonder, more joy, more risk, more suffering—that we see in every infant. Whatever we say about the mind (including what I say here) becomes one more idea within it capable of being debated and negated. Nothing so characterizes mentality as its inherent slipperiness. It is the shape-shifter supreme, always on its way toward a new identity. Hence, it cannot encompass itself. That very paradox is an evolutionary one. It is grounded in the fact that at a certain point, evolution reaches a reflexive state that generates the idea of evolution.

Over the past two generations, evolution has become the most comprehensive scientific concept since Newton's laws of motion. Beyond living things, it is now invoked to explain the emergence of matter out of the big bang—the spontaneous organization of prebiotic molecules, the development of stars and galaxies. The human mind, which reaches out to grasp the cosmic process from which it has emerged, clearly holds a special, frontier position in evolution. But it is not any one focus or fascination of the mind that points the way forward; it is the whole mind (or as much of it as any of us can experience) exercised in a condition of graceful integration.

There are forms of mysticism, like Zen Buddhism, that use an open, nondiscriminating style of meditation meant to bring us close to appreciating the expansiveness of the mind. The impish humor of Zen stems from the ability of the mind to stymie itself with paradox—and become larger by that act. Ultimate mind is that which appreciates the paradox, though unable to explain it. Thus, the Zen masters spoke of the mind as "a sword that cuts but cannot cut itself, an eye that sees but cannot see itself." It may not be beyond computer science to find the same wise delight in the mind's often comic effort to capture itself.

Perhaps, then, with a bit of humility and a sense of humor, computer science can help us learn something about the mind's radically transcendent role in evolution. After all, it is the human mind that invented artificial ones (as much for the fun as for the utility of it) and still has room left over to defy the logic or grow bored with their predictable correctness. Left to their own devices, can one imagine computers creating dada art or the theater of the absurd? Transcendent context is the evolutionary margin of life still waiting to be explored. What computers can do represents so many routinized mental functions we can now delegate and slough off as we move forward to new ground. The machines are behind us, not ahead.

Notes

1. Alfred Russel Wallace, *Darwinism,* 3rd ed. (London: Macmillan, 1905), 463.
2. Jonathan Howard, *Darwin* (New York: Hill and Wang, 1982), 65.
3. Steven Pinker, *How the Mind Works* (New York: Norton, 1997); Daniel Dennett, *Darwin's Dangerous Idea* (New York: Simon & Schuster, 1996).
4. Carl Sagan, *The Dragons of Eden: Speculations on the Evolution of Human Intelligence* (New York: Random House, 1977), 225.
5. Dennett, *Darwin's Dangerous Idea,* 370, 380.
6. Marvin Minsky, quoted in William Stockton, "Creating Computers that Think," *New York Times Magazine* (December 7, 1980): 41.
7. Robert Jastrow, "Toward an Intelligence Beyond Man's," *Time* 111 (February 20, 1978): 59.
8. Ray Kurzweil, *The Age of Spiritual Machines: When Computers Exceed Human Intelligence* (New York: Viking, 1999), 5.
9. Robert Jastrow, *The Enchanted Loom: Mind in the Universe* (New York: Simon and Schuster, 1984), 166–167.
10. Mark Slouka, *War of the Worlds: Cyberspace and the High-Tech Assault on Reality* (New York: Basic Books, 1995), 97.
11. Frank Tipler, *The Physics of Immortality: Modern Cosmology and the Resurrection of the Dead* (New York: Doubleday, 1994), 225.
12. Ibid., 242.
13. Jennifer Cobb Kreisberg, "A Globe, Clothing Itself with a Brain," *Wired* 3.06 (June 1995): 108–113.
14. Kevin Kelly, "Singular Visionary," *Wired* 3.06 (June 1995): 161.
15. Bill Joy, "Why the Future Doesn't Need Us," *Wired* 8.04 (April 2000): 244.
16. Jaron Lanier, "Agents of Alienation," *Journal of Consciousness Studies* 2.1 (1995): 76–81. Also see Lanier, "One-Half of a Manifesto," *Wired* (December 2000).

17. Howard, *Darwin,* 71.

18. Stephen Jay Gould, quoted in Dennett, *Darwin's Dangerous Idea,* 390.

19. Erich Jantsch, *Design for Evolution: Self-Organization and Planning in the Life of Human Systems* (New York: George Braziller, Inc., 1975).

20. Henri Bergson, *The Two Sources of Morality and Religion,* trans. R. Ashley Audra (Notre Dame: Notre Dame University Press, 1977), 309.

21. Norbert Wieners, *The Human Use of Human Beings: Cybernetics and Society* (New York: Doubleday Anchor Books, 1954), 183.

Louis Feuillee, *A Monster Born of a Ewe.* Courtesy of the National Oceanic and Atmospheric Administration (NOAA) Photo Library

Letter from Charles Darwin to His Sister, Catherine

Simmons B. Buntin

Letter No. 1

21 January, 1832
My Dearest Catherine,

Passage to the Cape Verde Islands,
a minor stopover for the *Beagle*,
but a major one for myself.
Oh, if you could have seen my face—
the color of stitched linen at Downs
(where last I have seen either you or Susan).
How can I explain my misery at that time?
The tormenting waves, the incessant rocking,
always rising and collapsing
as my stomach did the same.
Fitzroy is a fine man,
as he would look in on me while
I lay idle at sick bay;
But Wickham, his first mate,
knew no friendship for me.
My quarters fare little better—
I share the poop cabin,

and have my drawers; the two others
(officers both) have lockers.

16 March, 1832
Finally it is Spring—
it seems as if even these vast seas
know the changes. They are richer,
though I knew well before we reached the mainland
we were there. A single leaf, a barkless twig,
a clod of saturated grass, still living—all signals.
No beauty exists in all the world
such as in these tropical lands.
In all my days of studying,
under Henslow or even Sir Adam Sedgwick,
I was never prepared for the absolute
numbers and grand diversity of life—
of species. I have been able to collect,
though I must have killed
hundreds of insects, small mammals, and birds.
(Do not worry, Catherine, I know how
you love life. These species are too numerous
for my sampling to harm.)

One butterfly must be named for you—
its wings are the majesty's blue blazoned
with scarlet, violet, and even silver.
How much it reminds me of your favorite brooch.
These lands have too many more to describe,
the brilliantly colored parrots, the gay
primates swinging on twisted branches . . .
Father must accuse me
of lizard-catching now, as well.

Yet in all of this beauty, one thing
remains disturbing. Here
on Bahia, on the Northeastern coast

of Brasil—chiseled into the delirious
greenness of rainforest—
man holds man captive.
Nothing plays enchanting in blood
mixing with sweat on the whip-cuts
of the negroes. Nothing enchanting
in the deep brown skin
chained with iron coils.
You must see the difference.
I collect a few specimens for knowledge,
for all—it is my passion, no man sees harm.
But these men, vulgar and cruel,
they act as if they transcend the Creator,
though He who created such solitudes
surely must not agree.

We depart for the South
in but a short while. I cannot say
I will be home soon—the *Beagle*
shelters my bed now, much as
the tropical canopy is secure in the mist.
You cannot know
unless you see these forests
and breathe this air . . .

With loving passage,
Charles

Letter No. 2

9 June, 1834
My Dearest Catherine,

Our course lays due south, a new passage
through the Strait
of Magellan, and I cannot fathom

what strange currents lurk
beneath the iron clouds. Once
I captured the alien
view of Southern glaciers:
inverted domes rimmed with purest
white (oh, how the stars must be jealous!);
but Catherine, it is their blue
which holds me.
Fitzroy remarked
these are the frozen flames of Vulcan,
though I questioned the atmosphere
and found other evidence: ice
crystals gathering and refracting
the light. A simple combination
of muted sky and sea.

Yet I fear this voyage
is leaving me too scientific—it is not
some chemical reaction or
ice cones permeated by tropospheric rays.
There is more; and
I can only say, when I see these glaciers,
I am reminded of mother's eyes.

Beneath heavy skies,
however, we are threatened
by harrowing winds and black
fingers of basalt.
These are unexplored waters,
so I am braced by the cartography, the geology—
yet I must fear
a wooden hull's limitations.

28 July, 1834
Valparaiso!
We have anchored
(both our wind-tattered sails

and our restless feet) at the chief
seaport of Chile, the city
whose fragrances recall the intricate
tropical gardens of St. Cruz in Tenerife.
And if the dense green
forests of Brasil cause your eyes
to ache, then Aconcagua
and the long chain of Andes
will leave you blind!

I am reminded again
of the numerous species
which make up the grandeur of life:
I have seen, in the high
hills of Patagonia,
a bird larger in wingspan
than a British skiff's sails, and more
buoyant. I have seen on the uneven
playas of Tierra del Fuego a dumb and
flightless bird six hands higher than my brow.
And I have seen, weaving
the icy Antarctic waters, a slick
bird whose wings
are more efficient
than the finest pair of fins. And I have found
a striking likeness in their thin bones,
in dry feathers . . .
Every evening I ask the Creator,
How long are the days of the Genesis,
oh Lord? Yet I cannot discuss
such a heresy with Fitzroy, who nearly abandons me
upon a lifeless rock in the Pacific;
but with you, I can leave
these questions, and more . . .

In loving passage,
Charles

Letter No. 3

9 October, 1835
My Dearest Catherine,

We have sailed from the anarchy
of Lima and Peru
for the drier anarchy of the Galapagos,
where volcanic craters burn
without lava—
their regular forms jutting
from the archipelago
like the great iron-foundries
at Staffordshire.
And though there are no plumes,
the slight vapour blends
with low sky so that once again
the world is gray.

It is gray in the mutinied captain's
skull found among salt-green
succulents, in the oppressive
heat of absent wind, and
in dusky hues of equatorial finches.
Perhaps it is my mood which is truly
gray, as Fitzroy turns
madder with the days
and crewmen yearn for British seas.

Yet we are here, among these
curious rocks, and surely there is hope
in their exploration.

25 October, 1835
What joy in the cloudless skies,
in these barren isles! Though I have found

few species, it is their rarity
which excites. On Albemarle,
the largest island, I have tossed
a remarkable lizard by tail into the sea.
And always he returns!
On Chatham Island
I have balanced unsteadily
upon the giant back of a tortoise grazing
the sweet red fruit of cactus!
And of thirteen species
of finch, where I was drowning
in the dullness of feather,
I now sail on the varied waves
of their beaks!
Come sail with me
Catherine—take the wind west
to these juvenile isles and dance
among the gray feathers
that make up the brilliance of life.
If I appear too drunk to write
with steady hand and level mind
it is because I am too
undernourished not to go on.
Though sailors laugh
as I sketch the remarkable shapes
flourished since just one finch
lit upon Indefatigable's jagged
beach, I am aware only of life's
ability to persevere,
and evolve.

But in man's own wilderness,
void of cottages and cobblestone
and into the saline deck
of navigator's ship, perseverance
usurps evolution, discarding it quite

entirely. No, you should not dance here.
Dare say that I should not, either—
but for these birds and vines
and islands. And the faint memory
of a distant home.

In loving passage,
Charles

Is Evolution a Social Construction?

Michael Ruse

In the middle of 1996, a well-known New York physicist, Alan Sokal, published an article with the quite remarkable title "Transgressing the boundaries: towards a transformative hermaeneutics of quantum gravity," in a leading journal of "cultural studies," *Social Text*. No sooner had the piece appeared than Sokal revealed that the article was a hoax, a spoof on the kinds of pseudo learning which he claimed appears so often in such journals. He, and his many supporters in the scientific community, believed that for too long now scholars in the humanities have been illicitly downgrading the status of science. Rather than appreciating science for what it truly is, namely a disinterested reflection of objective reality, such critics have been falsely labelling it a mere "construction," an epiphenomenon reflecting favourably on the society within which it is produced, with no more objective standing than a novel or philosophical thesis or religious sermon. Now, thanks to Sokal's prank, these naysayers have been revealed as pompous fools, with no true understanding or feel for that which they attack.

One would have to have a heart of exceptional purity not to smile at the discomfort of the editors of *Social Text*, who were so sure of the rightness of their cause that they did not bother to get Sokal's article refereed. But there are serious questions here about the nature of science, and evolution *malgré lui* is right in the thick of them. For twenty-five years now, critics have been arguing that evolution—the ideas and the theories—is the paradigmatic instance of cultural ideology masquerading as objective truth. It has been accused of epitomizing and legitimating all of society's darkest prejudices and practices: gross sexism, endorsing the power and status of men over women; appalling racism, favouring whites over blacks and gentiles over Jews; and promotion of extreme right-wing politics and economics: the kind of stringent libertarianism which would make Margaret Thatcher feel "positively wet." Nor are the critics to be found only in faculties of humanities: feminists, social scientists, philosophers, and the like. Even evolutionists themselves throw a cynical light on their work. Stephen Jay Gould begins one of his books by admitting that "change through time does not record a closer approach to absolute truth," opting rather for a view of evolution where what counts are the "cultural contexts that influence it so strongly." And Richard Lewontin, than which there is no greater, writes recently: "Darwinism, born in ideological struggle, has never escaped from an intimate reciprocal relationship with world views exported from and imported into the science."

Polemics apart, what is the truth? Is evolutionary thought about a real world, a product which is independent of the producer, or is it simply a reflection of the scientists and their times, an advertisement for societal codes and practices? The best way to answer this question is to cast it as a debate or difference about norms or values. If evolution (evolutionary thought, that is) is objective, about reality, then it should be subject to those rules which scientists cherish because they are thought to take us beyond the immediate and subjective: it should be guided by values or norms which take us right across to the independent truth. These norms, often called "epistemic values" (from the Greek word for knowledge), include such things as predictive accuracy, consistency both within the science and with other branches of science, unificatory power bringing many different areas under one or a few hypotheses, simplicity, and (perhaps most important of all) a kind of fertility—an onward gaze leading to new problems with hints towards their solutions. Something which exhibits these sorts of virtues does so because it transcends its origins. In the memorable words of Karl Popper, it is: "knowledge without a knower." If to the contrary evolution is subjective, then its masters are cultural norms or values based on such things as religion or sex or class or politics. Evolution simply reflects the personal preferences of individual scientists or the assumptions and beliefs of the society in which their science is produced. It is knowledge where the knower—man or woman, black or white, Jew or gentile, socialist or capitalist—is all-important.

Now, if we look at the history of evolutionary theorizing, starting with its origins back at the end of the eighteenth and beginning of the nineteenth century, with people like Erasmus Darwin (Charles's grandfather) in England and Jean Baptiste de Lamarck in France, it cannot be denied that its *raison d'etre* was their cultural norms and aspirations rather than anything particularly epistemic. These men were indifferent to such things as predictive excellence, and hardly more enthused about perceived consistency with other sciences. They did care very much however about the dominant social ideology of the day, progress: improvement of knowledge and social conditions through human effort. Consequently, they saw their evolution as a vehicle for and justification of such progress: they saw evolution as an upward progression from the blob, the "monad," to the apotheosis, the "man." In good circular fashion, these evolutionists thought the social justified the biological and the biological justified the social. (People often say that cultural evolution is Lamarckian, because it centres on the transmission of acquired characters. But that is the point. Biological Lamarckism was cultural.)

But before you think that this was an end to things, that a pattern was set never to be changed, you should know also that the evolutionists' contemporaries—notably the great comparative anatomist Georges Cuvier—were absolutely contemptuous of the evolutionists. This contempt came not because Darwin and Lamarck were evolutionists *per se*, but because they violated so blatantly the then highly regarded norms of good science: precisely those norms listed above, namely predictive power, consistency, fertility, and the like. Evolution was judged a pseudo-science simply because it was so obviously no more than a cultural construct, an excuse for the promotion of social and similar values. Not that one should conclude that Cuvier himself was simon-pure in his criticism. Indeed, it was thoroughly self-serving. This was less because he had his own cultural commitments, than because as a Protestant making his way in a very conservative Catholic France, it was a crucial part of Cuvier's career strategy to argue that science is the one area where—by virtue of its cultural-value neutrality—the personal religious and other social commitments of the practitioner are irrelevant!

Charles Darwin of course altered things in many ways, thanks both to the strong case he made for the very fact of evolution as well as for his mechanism of natural selection: the force he proposed as lying behind organic change. After the *Origin*, evolution was no longer a pseudo-science: if nothing else, Darwin's unification beneath the evolutionary umbrella of so many branches of biological inquiry—instinct, palaeontology, biogeography, anatomy, systematics, embryology—meant that the epistemic status of the field was lifted to levels far above its early state. But Darwin still wanted to have his cultural commitments right there in his science. In the *Descent of Man* especially, Darwin showed that he held, and thought his evolutionism supported, some very conventional Victorian beliefs about the superiority of the white races, the virtues of capitalism, and the strong bold intelligent nature of men and the soft sympathetic emotional nature of women. There are times when you think you are reading a novel by Dickens, rather than a serious scientific treatise.

Again, however, one must take care with one's interpretations and conclusions. Virtually everyone became an evolutionist, but the field was never regarded as top-quality science, like physics or even physiology. By and large, it was not a university subject, but rather more a kind of pop science, suited for the museum and public displays, or for the lecture hall especially in front of a general audience, or for the nonspecialist magazine or newspaper. In part, this failure of evolution to make the top grade was purely a function of finances. Men like Thomas Henry Huxley, for all that he took on the label of "Darwin's bulldog," had a cold eye when it came to the cash. They could see possibilities for support for themselves and their students from certain subjects. Physiology, for instance, they sold to the medical profession as an essential part of any future doctor's training. Evolution did not seem to offer such possibilities and so there was little pressure to remove the cultural and stress the epistemic.

However, in part, the keeping of evolution at the popular level was deliberate. Huxley and friends were fighting the establishment, an establishment founded on the Anglican Church, and they wanted their own alternative. It suited them to use evolution as a kind of secular religion, one which gave a full world picture to challenge Christianity, even to the extent of providing a moral code: Social Darwinism. Part as cause and part as effect, because Darwin's fellow Englishman Herbert Spencer was truly eager to promote the cultural at the expense of all else, we find that in many respects it is his evolutionary thought which became most significantly and most widely influential as the nineteenth century drew to a close. It is well known how American barons of industry—John D. Rockefeller the First and Andrew Carnegie—became ardent Spencerians; but, even as far away as China, Spencer's philosophy became *de rigueur* for the forward-looking young intellectual.

Things changed again in the 1930s and 1940s. The mathematicians—R. A. Fisher, J. B. S. Haldane, and Sewall Wright—showed how Darwinian selection and Mendelian genetics are complements, and then the naturalists—E. B. Ford especially in England, and Theodosius Dobzhansky and his colleagues in America—put the empirical flesh on the theoretical skeleton. Evolution became a much more professional subject, like physiology and embryology, and it moved into the universities and research laboratories. One finds that there was a much more serious effort to produce science which is predictive, over small numbers of generations if not over millions of years; a science which is consistent with other areas of inquiry; a science which is forward looking, producing research programmes for young ambitious scientists; and much more. At the same time—shades of Cuvier—there was realization that evolution's status as a pop science or secular religion was holding it back, and there was conscious effort to keep cultural values at bay. Julian Huxley, grandson of Thomas Henry and an ardent evolutionist in his own right, but also one who thought that evolution is the answer to all of life's mysteries—he wrote a book on evolution entitled *Religion Without Revelation*—was admired and praised, and then kept firmly outside the circle! He was denied grants, and there was to be no nonsense about letting him edit the newly founded professional journal *Evolution.*

One might add, perhaps one could have predicted, that these new professional evolutionists—founders of what they called "Neo-Darwinism" or the "Synthetic Theory of Evolution"—tended themselves to have all sorts of cultural values, especially an obsession with progress, which they thought part and parcel of evolutionary thought. But they realized that too obvious a parade of these values would be antithetical to their aims. So they set about writing two sets of books. One purely professional, presenting mature science, with supposedly no cultural values. Then another, with the mathematics gutted (not much work here!), explicitly addressed to the "general reader," and loaded with cultural values about progress and democracy and anti-communism and so on and so forth. The paleontologist of the group, G. G. Simpson, was the paradigm. His *Tempo and Mode in Evolution* (1944) was sternly professional and eschewed all talk of progress and like topics. His *The Meaning of Evolution* (1949) gave away its popular intent in its subtitle: "A Study of the History of Life and of its Significance for Man." And then, in 1953, we are back with a revision of *Tempo,* and nary a wisp of a cultural value entered into *The Major Features of Evolution.*

Today, if you look through the pages of *Evolution* or of the *American Naturalist* or of comparable journals, you find that the synthetic theorists, whatever their motivations and their own beliefs, did their work well. It simply is not the case that modern professional evolutionists parade cultural values, although they certainly do take very seriously values thought conducive to truth—predictive fertility and the like. Even where the temptation might be there, it is avoided. Nicholas Davies's study of the dunnocks, where he recorded monogamy, polygyny, polyandry, and even polygynandry (the polite name for group sex), would seem the perfect vehicle from which to launch into an endorsement of the *Playboy* philosophy. But one searches in vain. The scientist might be a libertine or he might be a prude, but you will never tell from his discussions of his birds.

It is significant that often, too often, those who want to find cultural values in evolutionary writings turn either to the past—which we have just seen is not like the present—or to popular writings, those of Dawkins or Gould himself, where even today cultural values are allowed and perhaps expected. Dawkins on God, for instance, or Gould on baseball. But given what we now know of history does this mean that we can now simply dismiss these "deconstructivists"? At one level, I think we can. The critics really are ignorant of evolution and its history. Although do note that we do have a kind of funny double-layered causal factor at work here, with values of one sort crowding out values of another sort. Because scientists, evolutionists in particular, value the status and effects and benefits of being professional scientists—that is, of being scientists who will have naught to do with cultural values—they push such values down and out of their mature, professional science. It is not that they do not hold such values, or even are glad that such values are not in science, but that they realize that by incorporating such values they cannot realize other aims they hold more highly. (If you think there is never a tension here, look at the difficulty that Edward O. Wilson sometimes has in keeping his many social and religious views out of his science.)

Yet this is not quite all. Today's evolutionists do not use their science as a medium to promote their social and cultural opinions and likes and dislikes. But this hardly means that their science is culture free. The values may not be there but the culture remains. Especially through the great use that scientists, including evolutionists, make of metaphor, culture was, is, and surely always will be, a major component of science. And by culture, I mean precisely the ideas and theories and practices of groups of humans living together in a civilized—"cultural"—fashion. Even the briefest glance at contemporary evolutionary biology, with all of the talk about "strategies" and "arms races" and "optimality models," not to mention "competition" and "selection" and much more, ought to flag you to this. But let me make my point with one of the oldest and most modern of evolutionary metaphors: "the division of labour."

Introduced and highlighted by Adam Smith in the eighteenth century, it was at once picked up and used by Erasmus Darwin. But it really came into its biological own in the nineteenth century thanks to the Belgian-born Henri Milne-Edwards, who used the idea to explain the diversity and working of bodily parts. It was embraced enthusiastically by Charles Darwin, who saw it promoted by natural selection, both within the individual and its parts, and within groups and their members. Moreover, the metaphor was used approvingly by Darwin, suggesting that its manifestation and refinement is a good thing, a sign of progress, culminating in humankind. Such an endorsement is precisely what one would expect from the grandson of Josiah Wedgwood, who made his fortune by applying the division of labour to the pottery trade in the late eighteenth century British Midlands.

Today the metaphor is used even more than ever, but the values are gone. E. O. Wilson, for instance, has made much of the division and its importance in his path-breaking studies of the social insects, and his analyses of why the females break down into as many castes as they do: workers of various kinds and sizes, with different functions inside and outside the nests, not to mention the soldiers and their variations. Using the metaphor he is able to show that the different forms, morphs, fit exactly into patterns or categories suggested by the optimal effects of natural selection. Yet for all that elsewhere he may be given to reading his values into his science, this is one point where he simply does not do so. Whether Wilson thinks that the division among humans is a good thing, because it promotes efficiency, or a bad thing, because it deadens the soul, among his ants Wilson is entirely neutral. They do not care and think about what they are doing. They are simply programmed to act by their genes and that is it. End of the matter.

The point however is that Wilson, in common with other of today's evolutionists, is structuring and understanding his biological world through the lens of ideas taken from his culture. He simply could not have done what he is doing had he lived in the sixteenth century or had he been brought up among the Yanamamo, and never had the influence of an industrial society and its practices and philosophies. And before you retort with the response, which incidentally goes back to Aristotle, that although Wilson uses metaphors they are in principle eliminable (and hence goes the culture), let me point out that neither Wilson nor any other evolutionist shows any inclination to eliminate their metaphors. If they did, they would cut themselves off from one of the most prized epistemic virtues of good science: its predictive fertility. Good metaphors like the division of labour are heuristically powerful, pointing scientists in new directions. No sensible scientist is going to geld his or her theories for the sake of philosophical purity.

So what is the conclusion? Is it ultimately that evolutionary theorizing is subjective, a construction guided by values or ideology, through and through? Not at all!

It is cultural, but that is another matter. Good science, including good evolutionary science, is governed by disinterested norms promoting the virtues which are thought indicative of truth, of knowledge about the world if you will. The point is that we do this through the lens or the filter of our culture—where lenses and filters are thought of as good things and not simply blurring the picture. But this is hardly a surprise because in a sense this filtering is true of everything. I see the Eiffel tower, but where do I see it from? Underneath? The top of Notre Dame? Or wherever? Different views, but the same thing. And that is not even to take into account that I see the tower with my primate organs: clearly, in colour, and so forth, without echo location and the like. All knowledge is in a sense an interaction between the world as we suppose it and the perceiving subject. This is no less true of evolutionary knowledge. The point is that science, and this is true of today's evolutionism, rises above the purely subjective, bringing order and imposing standards, and gaining results: prediction, unification, consistency, and more. No matter what the critics say, that is no small or modest achievement.

Suzanne Stryk, *Afterimage, (Regal Moth)*

Science Evolving: The Case of the Peppered Moth

Craig Holdrege

The peppered moth is used in high school and college biology courses, as well as in many textbooks, to illustrate evolution by natural selection. The textbook story goes like this:

The peppered moth (*Biston betularia*) is a light-colored nocturnal moth. In 1848, a dark (melanic) specimen of this species was discovered near Manchester, England. In the latter half of the nineteenth century, the frequency of dark peppered moths increased, especially in woods around industrial centers. By the 1950s, scientists found only the dark form. In these woods, why does this evolution of the peppered moth from light to dark pigmentation occur?

There's a simple answer. As the pollution in the forests around industrial areas became worse, the light-colored lichens on the trees died, and soot darkened the tree trunks. The light-colored moths were no longer camouflaged, and the rare dark mutant now had a selective advantage. Birds that preyed on the moths couldn't see the ones with dark wings. Those moths thus had the opportunity to reproduce, allowing their populations to grow rapidly, as the populations of conspicuous white moths were being decimated.

Since the 1960s, clean air acts in Great Britain and the United States have led to markedly improved air quality around industrial centers, and the numbers of dark moths have fallen significantly in the forests, while the light moth is again becoming more prevalent. The evolutionary trend is reversing. A striking case of parallel evolution has been found in the forests near Detroit, which had over 90 percent dark moths in 1960 but in 1994 only 20 percent.

This type of evolution has been called industrial melanism. (Melanin is the pigment that makes the wings dark.) It exemplifies the Darwinian view of evolution: A species displays phenotypic variation (light and dark forms) on which natural selection can operate. In this case birds selectively feed on conspicuous moths, and because the background coloring changes, the moth evolves. As Bernard Kettlewell, the Oxford biologist whose research made the peppered moth into a textbook example, summarized, "Had Darwin observed industrial melanism he would have seen evolution occurring not in thousands of years but in thousands of days—well within his lifetime. He would have witnessed the consummation and confirmation of his life's work."[1]

In the early 1980s I began teaching about peppered moth evolution in a university preparatory, high school biology course in Germany. Using this example, I could clearly develop the concepts of mutation and directed natural selection as factors of evolution, concepts required in the state-regulated curriculum. Since I was teaching the peppered moth only as an example to make certain concepts clear and could spend only a short time with this theme, I used textbook descriptions and other secondary sources. Essentially, I taught the same story described above.

In 1986 I came across a short report on new research concerning the peppered moth that gave me an awakening jolt. The last sentence stated that Cyril Clarke, a British scientist, had investigated the peppered moth for twenty-five years and in that time had seen only two specimens in their natural habitat during daylight.[2] *What is going on here?* I asked myself. I'd been showing students photographs of the moths on tree trunks and telling them about birds' selectively picking off the conspicuous ones. And now someone who had researched the moth for twenty-five years reports having seen only two moths. I immediately ordered Clarke's article, and my study of the primary literature began.

As strange as it may seem, no one knows where the peppered moth lives during the day. Some have been found on the underside of branches or in the shadow beneath the angle where a branch meets the trunk. It's clear that the one place they are not resting on is exposed tree trunks. How, then, have the moths been studied? Researchers enter the forests at night and turn on bright lamps that attract nocturnal insects. In this way, they capture the moths. They also set up so-called assembling traps housing virgin females that release pheromones into the air, attracting males into the traps. The males fly into the assembling traps only at night; they are never caught during the day. Since one rarely, if ever, sees these moths during the

Bruce S. Grant, *Biston betularia cognataria,* (the peppered moth). The pale form (top) and the melanic form (bottom). Courtesy of American Genetic Association

day, it is assumed they are resting somewhere in the forest, becoming active at night.

If the moths aren't observed during the day, where do the beautiful photographs of the moths on trees come from? In general, authors don't report the conditions under which the photos were made. I have found references only in a 1975 article written by the British moth researchers David Lees and Robert Creed. The piece describes how the moths are killed, glued to the tree surfaces, and then photographed.[3] Most photos in textbooks are reprints taken from Kettlewell's work and give a striking image of camouflage—against dark bark, a dark moth is virtually invisible, while a light moth sticks out like a sore thumb. Since Kettlewell worked with live specimens, he must have placed them intentionally to show camouflage or lack of it.

Unsuspecting readers will normally (and perhaps naively) assume, unless otherwise informed, that they are looking at a natural phenomenon. The impressive image of camouflage in the peppered moth sticks in the mind, especially when the image is accompanied by a brief tale, detailing how they've evolved, a tale that gives no hint that we are looking at an artificially constructed situation. And as most textbooks state, the explanation of industrial melanism appears in view of such images almost "self-evident." This self-evident explanation dissolves when we learn that researchers can't find the moth during the day and that the pictures are composed by photographers.

In the 1950s, Kettlewell undertook a series of impressive experiments to see if he could observe experimentally what nature might be doing in a more hidden way. He bred moths in the laboratory in order to have large enough numbers for experiments, especially females, which rarely flew into the light traps at night. He then marked the moths on the underside of the wings for later identification. The light and dark forms of the moths were then released early in the morning into unpolluted and polluted forests. He later recaptured some of the moths in the light and assembling traps. Kettlewell summarizes the results of two such mark-release-recapture experiments: "In an unpolluted forest we released 984 moths: 488 dark and 496 light. We recaptured 34 dark and 62 light, indicating that in these woods the light form had a clear advantage over the dark. We then repeated the experiment in the polluted Birmingham woods, releasing 630 moths: 493 dark and 137 light. The result of the first experiment was completely reversed; we recaptured proportionately twice as many of the dark form as of the light."[4]

There is a clear correlation: In polluted forests, more dark moths are recaptured, and in unpolluted forests more light moths are recaptured. But the experiments do not reveal whether birds are feeding on the moths. Kettlewell investigated this question by performing other experiments. In collaboration with the well-known Dutch ethologist Niko Tinbergen, Kettlewell released moths (not for recapture) onto tree trunks, where the moths remained stationary. The scientists hid and observed birds feeding on the moths; Tinbergen even filmed the process. Generally, the more conspicuous moths—those on the "wrong" background according to our human standard—were taken first, and after all the conspicuous moths were eaten, their numbers were replenished. Camouflaged moths were also eaten, but not as many.

In an aviary, Kettlewell made similar observations. A pair of birds, great tits, took no moths within the first two hours, but then within an hour, they had eaten most of the conspicuous moths and a few of the camouflaged ones. The second time the experiment was performed, all the moths were taken much more quickly, within one-half hour of being released. "It was suggestive that the tits were becoming specialists on *betularia* [the peppered moth], and subsequently they were seen to be searching each tree trunk eagerly one at a time immediately after admission, thereby defeating the purpose of the experiment."[5]

Kettlewell believed his research proved that the evolution of the peppered moth is caused by selective predation by birds. But how compelling is this conclusion? Consider the results of his aviary experiments. He observed that the two birds were much quicker at taking moths after they had had experience in doing so. They found the camouflaged moths as well. If one takes this experimental evidence and imagines it transferred into a natural habitat, wouldn't it be reasonable to think that as the dark peppered moth began to spread initially (for some unknown reason), the birds might have begun to recognize them as well? This conclusion is just as sound, but also just as speculative, as Kettlewell's, which states that since the birds feed on the conspicuous (light) form first, its numbers have decreased while the dark form has increased its numbers.

The reduction in the lichen covering of trees, due to air pollution, in forests around industrial centers has been viewed as a primary factor in the evolution of the peppered moth, since fewer lichen would make the light form more conspicuous and the dark form better camouflaged. In forests near Liverpool, the proportion of dark moths was over 90 percent in 1959, while in 1984 there were only 61 percent dark moths; the population of light moths has been making a dramatic comeback. The air pollution has decreased in this time, falling steadily from 1962 to 1974, and has remained since then at a constant low value. Although green species of lichen have repopulated trees, the light species of lichen, on which the light peppered moth is so well camouflaged, is still absent in the forests. Similarly, in forests near Detroit, the light moths increased from under 10 percent of the population in 1960 to over 80 percent in 1994, even though the lichen flora did not change perceptibly in this period. As a result, American biologist Bruce Grant believes that the role of lichens in the evolution of the peppered moth has been "inappropriately emphasized."[6] Clearly, if the lichen abundance has not changed, then it is very difficult to understand how selective predation by birds could be the primary factor in the evolution of the moth forms.

This is not the only feature that contributes to the dissolution of the clear-cut textbook story. In 1975, David Lees and Robert Creed conducted research in rural eastern England.[7] With certain variations, they basically repeated Kettlewell's experiments. In these forests, there was little atmospheric pollution, and the bark of the trees was "relatively light." When they glued dark and light dead moths onto trees, human observers found the light form better camouflaged than the dark form. They came back to the trees at regular intervals and counted how many specimens of each type of moth was still present and how many had disappeared, presumably having been eaten by birds. The results fit well with the observation of conspicuousness: more of the better camouflaged light moths remained longer on the trees than the more conspicuous dark moths. When, however, Lees and Creed captured wild moths in traps, there were about 80 percent dark moths and 20 percent light moths—exactly the reverse of what would be expected on the basis of the experiments. If resting moths are hunted by birds during the day, then the light form would seem to be at a selective advantage. Yet the forests seem to be populated by many more dark moths than light moths.

In 1979, Stephen Jay Gould and his Harvard colleague Richard Lewontin wrote a decisive critique of what they call the "adaptationist programme," characterizing "its unwillingness to consider alternatives to adaptive stories."[8] This "unwillingness" stems from a preformed idea that has the quality of a conviction. The idea that selective predation by birds is the primary causative factor in the evolution of the peppered moth became a fairly rigid paradigmatic framework under which all facts have been subsumed. If Kettlewell hadn't been so convinced of the truth of bird predation causing peppered moth evolution, he might have left more room for alternative explanations.

In peppered moth research, we see how strongly a theoretical framework informs the interpretation of the facts. When scientists have, as Lynn Margulis puts it, "an uncritical acceptance of the mesmerizing concept of adaptation," there is a real danger of seeing what one believes.[9] If this happens, then we get the oversimplified portrayals that turn science into dogma. It is not very difficult to show that natural selection is at work when one tacitly weaves the theory into the description of the phenomena. You get out what you put in. Rudolf Steiner, in 1886, already saw this fundamental danger within science: "The basic error of many scientific strivings today is that they believe they are reporting pure experience, while actually they are only reading out of experience the concepts that they already placed into it."[10] His words have not ceased to be true.

In 2002, a new book, *Of Moths and Men,* written by Judith Hooper, appeared on the peppered moth story. Hooper shows in great detail how scientists were on a mission to prove the efficacy of natural selection in their peppered moth research.[11] Although in recent years, articles critical of oversimplifying the results of peppered moth research have been published in scientific journals, the idea of natural selection has a very strong hold on the scientific mind. This is well illustrated by two negative reviews of Hooper's book in *Nature* and *Science*.[12]

The reviewers—both evolutionary scientists and one a peppered moth researcher—take Hooper to task on various issues, but one criticism that both reviewers expressed interested me most. They stated that Hooper was failing to make an essential distinction: although the peppered moth example may not show the exact *agent* (mechanism) that causes the observed changes, it clearly does show the *fact* of evolution through natural selection.

The problem is that the concept of selection includes at its very core a selective agent. You can't speak about selection without there being something doing the selecting. The "force" of natural selection is nothing more and nothing less than the specific agent(s) of selection; to speak of the force of natural selection in general terms is vacuous (and borders on the mystical). Until scientists know that predation by birds, the effects of air pollution on larvae, or other factors are at work, the concept of natural selection has no concrete content. Why, then, do both reviewers stress the fact of natural selection when no agent of selection has been clearly determined?

It's as if at all costs they want to save natural selection—as if losing it would cause their grip on evolution to disappear altogether. Natural selection is felt to be the best weapon against the threat of creationism, and if scientists don't have this weapon, then what? But evolution doesn't stand or fall with the idea of natural selection. Perhaps a narrow Darwinian interpretation does, but all the evidence of organisms changing over time can't just vanish. Both Darwinians and creationists are stuck in ideologies—fixed mental frameworks. Wouldn't it be much more in the spirit of scientific inquiry to leave different avenues of explanation open as one continues to investigate the phenomena? Scientists should enjoy uncertainty and discovery as much as they enjoy fitting facts into a neat conceptual framework.

If we are truly interested in understanding the phenomena and not mirroring our ideas in them, then we must become more aware of our thinking in order to make our mind a more adequate instrument of understanding. A basic but important realization is that in performing an experiment, we are creating a simple and relatively transparent situation that is, of course, not identical with the more complex system of interactions involved in any given natural phenomenon. We should be extremely wary of drawing conclusions that go beyond the experimental situation itself. Kettlewell's field experiments show that birds feed on moths released onto trees in the early morning. But since the moths are not normally found on lower tree trunks during the day, Kettlewell has created (as all experiments do) an artificial situation. We need to recognize this and not simply conclude that in nature, birds feed selectively on moths in the manner Kettlewell has shown. We need to hold back conclusions in order to free our thinking, to consider alternative explanations, and to realize what we do not yet know.

All experiments are guided by ideas. Without the concepts of natural selection and selective predation, most of the research concerning the peppered moth may well never have been performed. These ideas have guided and focused the research and helped scientists to formulate specific questions and discover new phenomena. Problems arise when we no longer handle a concept as an instrument to see more, but as something to be substantiated by nature. When we begin to view the phenomena selectively, seeing only what seems to confirm our theory, then the concepts that initially sharpened our attention begin to make us blind. If, in contrast, we can use hypotheses as a way to get started, well knowing that they need to be left behind when we confront the phenomena, then we begin to practice a flexibility of thought that leads us further into the complex richness of the phenomena, and not into a monolithic theoretical construct. We help science evolve.

In recent years, I taught the complicated picture of the peppered moth to high school seniors at the Hawthorne Valley School in upstate New York. This is an independent Waldorf School, and its curriculum is not state regulated. The students were fascinated by the peppered moth and the contrast between the simple story and the complex reality. We spent more time on this example than one usually would, because I wanted them to see how science actually proceeds as a process of discovery and transformation.

Teaching in this historical case study approach demands more classroom time and also more research on the part of the teacher than providing general overviews of material. But it brings alive science as a process. We learn how scientists make observations, formulate ideas and questions, and test their hypotheses through experiments. We see how contradictions arise, how concepts become rigid, and then—often in the face of resistance—how they are modified or even dropped. Students begin to think of science as an evolutionary process occurring in a historical context. What could be a more appropriate way to learn about the science of evolution?

By proceeding in this way, students gain knowledge, but their knowledge is dynamic, not static information. They develop capacities and ways of approaching phenomena that they can apply to a variety of situations throughout life. Young people are—if we have not corrupted them too much—open-minded and interested in the world. Certainly, it makes sense for them to learn science (and of course other disciplines) not as codified knowledge to be memorized but as a way of interacting with nature that leads to insights and also to ever-new questions.

A significant problem in the way science is taught, popularized, and in general filtered down into the minds of children is that students are filled with scientific dogmas. They "know" that in evolution, the fittest survive; they "know" that the brain is a computer; they "know" that the heart is a pump; they "know" that genes determine heredity. One task of secondary and undergraduate science courses could be to dissolve such dogmatic "knowledge," which in reality is only acquired opinion, by showing how science evolves. In a given course, one can do this for only a limited number of examples, but it is much more stimulating for students than imbibing and memorizing large amounts of noncontextual information.

Teaching science as process would mean either reducing the use of textbooks or transforming textbooks into compendia of case studies. In perusing textbook presentations of the peppered moth, I was delighted to find one biology textbook with a short description of the peppered moth in the section on evolution but under the heading "Biology in Process." The author describes Kettlewell's work briefly and then goes on to state that recent experiments raise doubts about the selective predation explanation. He thus calls attention to the unresolved questions.[13]

The American Association for the Advancement of Science has published Benchmarks for Science Literacy. It is part of the Project 2061 (the year in which Halley's comet will return; the project began in 1985, the last time Halley's was here), which has the purpose of helping to "transform the nation's school system so that all students become well educated in science, mathematics, and technology." Concerning scientific inquiry, the text states that high school students should learn that "no matter how well one theory fits observations, a new theory might fit them just as well or better, or might fit a wider range of observations. In science, the testing, revising, and occasional discarding of theories, new and old, never ends." Most of the book, however, stands in contrast to this description of science as a process. In the main body of the book, one finds for all grade levels the "benchmarks" for what should be known in a given field at that age level. In this way, the book emphasizes content, not process. For example, by the end of twelfth grade, students should know that "the theory of natural selection provides a scientific explanation for the history of life on earth as depicted in the fossil record and in the similarities evident within the diversity of existing organisms."[14]

Once we have learned that one of the most cited examples of natural selection turns out to be very unclear, doesn't this statement seem dogmatic? If we are teaching dogma, it is important to be able to know that natural selection is an explanation; if we are interested in giving a sense for the nature of the scientific endeavor, then it is much more essential to know how the concept is used, what it reveals, and what it doesn't reveal. Without intending it, this book gives a very good picture of a codified view of the nature of things. The conservative slant is discernible when the authors say, "It is important not to overdo the 'science always changes' theme, since the main body of scientific knowledge is very stable and grows by being corrected slowly and having its boundaries extended gradually."[15] If this "stable" body of knowledge entails the myriad phenomena that scientists discover, such as the fact that insects have six legs, then I can agree with this statement. But if scientific theories are included, then I think we should stimulate our students to continually question popular, established standpoints. We should stimulate continual scientific revolution. Just as in the Middle Ages it seemed to many a self-evident fact that the Earth was the center of the universe, most certainly many of today's "scientific truths" will become historical belief systems in the eyes of future humanity.

Once we break out of the strictures of fixed explanatory patterns, we can turn more openly to the natural phenomena themselves. If nothing else, the history of peppered moth research shows the need for very basic natural history, without which experiments and theories are anchorless. Many essential questions can be answered only by direct observation—as difficult as that may be.

Clearly, we need to know more about the life history of the peppered moth. Where does it rest during the day? What are its natural predators? How far can it fly? How long do the moths live? Similarly, a greater knowledge is needed about the egg, larval, and pupal stages.

At the same time, alternative interpretations for melanism in the peppered moth need to be actively pursued. Might melanism have completely other functions than camouflage, like increasing warmth absorption or structural stability in the wing? Or is perhaps melanism in the adult a secondary effect of differences in the larval stages? Some research suggests, for example, that larvae of differing genetic types may not have the same viability. Theodore Sargent of the University of Massachusetts, working with a different species of nocturnal moth, has found evidence that the plants that larvae feed on may induce or repress the expression of melanism in adult moths.[16]

There are certainly other possible interpretations of melanism in the peppered moth. I doubt that any one explanation will turn out to be the right one, since in the long run all biological phenomena show themselves to be interconnected with an array of factors. We should probably also expect that in different localities at different times, different explanations may be necessary. This is certainly not a comfortable situation if we are looking for the one cause of industrial melanism, but why should reality be concerned about our predilection for monocasuality?

One difficulty in our approach to the peppered moth is that we've studied it only as an example of evolution. We have not yet set out to understand the moth in its own right. From the outset, we have considered the moth from a limited perspective. It was interesting that a student of mine questioned whether the peppered moth really is such a good example of evolution. He said that what the peppered moth is really showing us is how a species, by having different forms, is more flexible and able to survive as *one* species; the populations and varieties of the species fluctuate, but the species as a whole continues to thrive. In this view, the peppered moth shows the adaptability, and not the evolution, of a species.

Such a shift in vantage point accentuates how the peppered moth has been reduced to a mere example of a general theory. Certainly, the moth became more and more of a riddle even within the evolutionary perspective, but it is important to be aware of the limitations of understanding implicit in how one formulates a theme. Since limiting is also a way of focusing and finding an entryway into a theme, we can't just abandon points of view. But what we can do is to take different approaches in different contexts to show that there are various avenues of understanding, each with its strengths but also its limitations.

For decades, the peppered moth has been a standard classroom and textbook example of evolution. Millions of students have learned this "living proof" of natural selection. The story they have been and are being told is most likely false or, to put it more mildly, filled with half-truths. This is not because teachers and writers are intentionally lying, or hiding and bending facts, but because the example is brought only to prove a point, so that complications appear extraneous to the argument (if not to the truth). Moreover, the idea of natural selection has become so ingrained in the modern mind that it can become like a pair of spectacles that one doesn't remove anymore. Concepts that become axiomatic deteriorate the complex and rich phenomena of nature into mere instances of overriding principles. Instead of illuminating, the idea becomes, in Goethe's words, a "lethal generality."

This tendency toward solidification is not what keeps science alive. Vitality in science comes from researchers doubting conclusions, making new observations, and constructing new experiments; from scientists thinking original ideas that break through the constrictions of dominant paradigms. Science teaching need not only serve the codified "body of knowledge." It can also serve ongoing exploration and the continual renewal of ideas. Since there is "more to melanism than meets the eye," peppered moth research can be an excellent teacher of the living scientific process.

Notes

1. Henry Bernard David Kettlewell, "Darwin's Missing Evidence," *Evolution and the Fossil Record, readings from Scientific America.* (San Francisco: W. H. Freeman, 1978), 33.
2. Cyril A. Clarke, "Evolution in Reverse: Clean Air and the Peppered Moth," *Biological Journal of the Linnean Society* 26 (1985), 189–199.
3. David R. Lees and Robert Creed, "Industrial Melanism in *Biston betularia:* The Role of Selective Predation," *Journal of Animal Ecology* 44 (1975): 67–83.
4. Kettlewell, "Darwin's Missing Evidence," 29.
5. Henry Bernard David Kettlewell, "Further Selection Experiments on Industrial Melanism in the Lepidoptera," *Heredity* 10 (1959): 296.
6. Bruce S. Grant et al. "Parallel Rise and Fall of Melanic Peppered Moths in America and Britain," *Journal of Heredity* 87 (1996): 351.
7. Lees and Creed, "Industrial Melanism in *Biston betularia,*" 66–83.
8. Stephen Jay Gould and Richard C. Lewontin, "The Spandrels of San Marco and the Panglossian Paradigm: A Critique of the Adaptationist Programme," *Proceedings of the Royal Society London* B205 (1979): 581.
9. Lynn Margulis and Dorian Sagan, *Slanted Truths* (New York: Springer-Verlag, 1997), 272.
10. Rudolf Steiner, *The Science of Knowing* (New York: Mercury Press, 1988), 31 (translated by Craig Holdrege).
11. Judith Hooper, *Of Moths and Men* (New York: W. W. Norton and Company, 2002).
12. Jerry A. Coyne, "Evolution Under Pressure," *Nature* 418 (2002): 19–20; Bruce S. Grant, "Sour Grapes of Wrath," *Science* 297 (2002): 940–941.

13. Albert Towle, *Modern Biology* (New York: Holt, Rinehart and Winston, 1989), 228–229.

14. American Association for the Advancement of Science, *Benchmarks for Science Literacy,* (New York: Oxford University Press, 1993), 8, 125.

15. Ibid., 5.

16. Theodore Sargent, “Industrial Melanism in Moths: A Review and Reassessment,” in *Adaptive Coloration in Invertebrates* (Proceedings of a Symposium Sponsored by the American Society of Zoologists, College Station, Texas, June 1990), 17–30.

Suzanne Stryk, *Reasons for Numbers* (detail)

Why Do Birds and Bees Do It?

David C. Geary

My third-floor faculty office has three rather large windows that overlook a nicely wooded park at the outer rim of the University of Missouri campus. At the end of a long day, I often spend a few moments—maybe more than a few on occasion—spying on the wildlife that resides in and among the trees. I'm not talking about the undergraduates, as they're a bit more subdued than the park's permanent residents. (This might not be true in other venues, but that's a topic for another day.)

I've grown fond of watching the many species of bird that do their business and make their living under my watchful eye. On one occasion, I was more than slightly amused to see a pair of sparrows converge from different directions and meet furtively in the tangle of leaves and branches of a nearby ash tree. On this day, their business involved a quick copulation, and then they were off again in different directions. No small talk or festive banter to top the occasion, just a few moments in the branches and then back to their respective social partners. What I had just witnessed biologists would call an *extra pair copulation,* or EPC, when individuals with different mates get together on the sly. I suppose other folks would call it something else, but a rose is a rose by any other name.

I can testify that birds do it and bees supposedly do it as well. I've never witnessed the latter and imagine that obtaining proof might require some type of covert and remote surveillance since they do it in flight—a sting operation, if you will. In any case, I'll take the word of entomologists who study such things.

Now, you don't actually need to be a scientist to realize and observe on occasion that all sorts of living things, most in fact, mate. Be careful. I'm not advocating that you peer through your neighbor's window on a moonless evening, as this type of voyeurism can result in all sorts of bad noise and an unscheduled visit from the local constable, not to mention some potentially unpleasant images (depending on the neighbor). What I am suggesting is that after a long day, you too might spend a little time outside paying particular attention to the teeming life flying above your head and squirming below your feet. You might run across a spider or, worse, a snake, but take the chance. Eventually you'll be rewarded with the sight of creatures, one pair or another, doing it. More likely than not, you'll run across some beetles. The world is full of them and, if you've ever seen these guys before, you might be able to imagine how interesting it would be to catch them in the act.

Perhaps you can come up with something a little more enticing. Don't get me wrong. I'm not pushing some sort of cross-species voyeuristic perversion as an avocation. The sight or even the thought of a dalliance between a pair of sex-starved beetles does nothing for me. The real curiosity is why all of these different creatures, from the lowly beetle to the smarter-than-average human, are doing it. When approached scientifically the question becomes, Why did sexual reproduction evolve? This is one of the most fundamental questions to confront the biological sciences since Darwin's and Wallace's independent discovery of the principles of natural selection. Of course, there has always been the necessity to reproduce. But why don't they, and we, reproduce in a more Dolly-like fashion—that is, make clones of ourselves? We don't have that option anymore, but way back in our evolutionary history, our very small, probably slimy, premammalian ancestors did. Scientists have suggested these creatures started doing it in order to have extra copies of DNA, in case one copy gets messed up, or to make us different from one another. But most scientists hardly think about the purpose of sex, although many, aside from the ones who have trouble maintaining eye contact and keeping up a conversation, probably get to participate in the act every once in awhile.

You might protest that life is just a lot more interesting and enjoyable when it includes having sex—or for some folks, the prospect of someday having sex. There are even entomologists out there who derive enjoyment (vocational, of course) from studying how beetles reproduce. Doing it is fun and so why even ask these silly questions? Why not just do it some more? Sex helps to keep our relationships together and increases intimacy, as well as entertaining us before a new sports season rolls around. However, all of these objections miss the point. Making clones of ourselves would not only save us the trouble of trying to find a weekend date and ultimately a mate; the banking of the next generation wouldn't require us to mix our genes with those of someone else. Mixing genes doesn't sound so bad, but it comes at the ultimate price of giving up half of yourself—that is, half of your genes—to make another person. You might say that when two individual people mix their genes to produce little ones, the result is two parents who will look after the well-being of their mutual progeny. It might also be a way to get some better traits for your children. When I originally got into the reproduction business, I thought, *There is no way in the world I want to try to raise a clone of myself.* We just wouldn't get along. So I found a wife, Leslie (married twenty years now), who is not only better looking than me; she is nicer, more sociable, and much more of a conformist than I am. Well, it worked for my son, Nick, but not my daughter, Corie. She's attractive, but ornery like her dad.

It is often true that two parents are typically better than one. This is the case with most species of bird, especially when the young have to be fed and protected for weeks or months. While mom stays home, dad flies off to get the grub, and they often switch roles. But for many species of gene-mixing creatures, only one sex (usually females) looks after the progeny once they are born or hatched (or whatever the case might be). This is how it works for most mammals, as well as insects, fish, and reptiles. In these species, the male who outduels all others—the king of the hill, the champion in the ring—tends to do it with lots of females while the other males are forced to sit back and watch. While the offspring are gestating in mom, the males go off and hold new competitions. Whoever claims the crown this time gets to mate with the females that are not yet pregnant, and thus a cycle is formed.

There are many other gene-mixing species, some species of fish and insect, in which neither parent looks after the progeny. The mother simply deposits the eggs in one of the local crevices and then she's off with the girls, or sometimes off doing it again with one or several of the boys.

Mixing it up, so to speak, must then have some very deep benefit, one that goes beyond sharing the responsibilities of parenthood or simply having a good time. Nick, my son who's now thirteen, and I were discussing this one morning not too long ago. After some blushing and hesitating, he argued that people just need to be a little different, or things would get pretty boring. "Well," I suggested, "doesn't that depend on exactly who does the mixing and matching, or should it be matching and then mixing?" I asked him to think about it some more.

Bill Hamilton, one of the greatest biologists of the twentieth century, began thinking it over several decades ago and did so for many long hours and many long years. In addition to discovering why sex evolved, Bill also figured out why bees and other very social insects work together so well (they are all highly related sisters), and among other accomplishments set the foundation, later elaborated by Bob Trivers, for understanding why unrelated individuals will often cooperate. Although I've read many of his scientific papers, I didn't know Bill personally, but a friend and colleague, Mark Flinn, did. Bill and Mark, in fact, shared an office during part of the time when Bill was most intensely focused on the evolution of sex. Mark tells me that Bill (who passed away recently) was gentle, unassuming, and brilliant. Like Darwin, he was among that rare breed who knows the right questions to ask and has the intensity and focus to find the right answers.

Bill's answer to the evolution of sex was pathogens—that is, things that make us sick. His point was that there are all sorts of nasty little critters out there—viruses, bacteria, and worms—that do their business and make their living at our expense. Viruses need our DNA and RNA to make copies of themselves, and many of these worms are headed for a place where the sun don't shine—warm and cozy for them, it seems. So what's a person to do? If we were in the Wild West having a shootout with these critters, they would beat us to the draw every time. Sure, we could and do build defenses to keep them in check (at least most of the time), but they reproduce so quickly that a few of them consistently find ways to break down our resistance. Once this happens, we get sick, and sometimes we die.

The answer to these nasty little critters is *sex*. (Maybe thinking about replicating viruses and intestinal worms doesn't get you in the mood, but perhaps it should!) Actually, sex won't help *you*, at least as far as these critters go, but reproducing sexually will help your children. By mixing your genes with those of someone else, you give your children an immune system that leaps over the tricks these critters use to get around. They eventually will catch up to the new defenses built by your children's immune systems, and thus the only way your children's children can outsmart viruses and bacteria is to reproduce sexually. It never ends. As the Red Queen told Alice in Lewis Carroll's *Alice in Wonderland*, you run and you run as fast as you can, but you never seem to get anywhere. This is the way life goes. You work and work but you never seem to solve the puzzle, and you can't stop, because once you do, shit really hits the fan.

Although sexual reproduction helps to stay one step ahead of viruses, it opens up a whole other can of worms, and this brings us back to those two-timing sparrows and the more general issues of dating and mating. The problem for us and all other gene-mixing creatures is that not all mates are created equal. Sexual reproduction not only benefits our immune systems; it makes us different and less boring, according to my son, in many other ways. Many of these differences are unimportant and uninteresting, but other things—intellect, looks, money, cars, or whatever it takes—help us and our children do well in life. This new issue involves finding the right mate, or mates depending on the species. Easier said than done, as many of you may know.

Charles Darwin was familiar with this difficulty—not just from his own experiences, but through careful observation of the sexes and their behavior. A natural-born naturalist, Darwin not only discovered how natural selection works; in 1871, he published the most important book ever written on mating and the different ways in which females and males approach the subject, *The Descent of Man, and Selection in Relation to Sex.* The subject was when, where, with whom, and why they do it. Scientists call this *sexual selection,* that is, the social behaviors involved in competing for mates and picking mates.

Just a few yards outside one of my three windows is an ash tree. It's so close, in fact, that the branches have grown above and around the window, which gives me a bird's-eye view, so to speak, of all that's going on within. And a lot is going on. This spring, there were three cardinals vying for this territory, huffing and puffing in the branches, trying to bluster and bully the others out of the tree. Darwin called this aspect of sexual selection *male-male competition.* The winner gets a spot from which he can sing to the ladies and strut his stuff, a place for wooing and nesting and, of course, mating. Here in Missouri, cardinals are called red birds, because during the mating season, the males shine a brilliant red, which drives the lady cardinals mad.

The twist comes from the fact that some males are a little redder than others. Indeed, some are just plain dull (not unlike some of the scientists I mentioned earlier). None of this would matter, except that the lady cardinals are extraordinarily picky. They're not going to do it with just any Tom, Dick, or Harry that glides along. They want the pretty males, a preference Darwin discovered through his careful observations of nature, terming this inclination *female choice,* another aspect of sexual selection. Female choice is not just picking the males that sport a colorful coat. Sometimes size *does* matter. Often, when looking for that special someone, female birds are searching for colorful males with large feather tails.

But why? This is where Charles Darwin meets Bill Hamilton. It turns out that these big and pretty males are pretty not just for the sake of aesthetics. They have the best immune systems. This makes perfect sense if one of the primary purposes (not the only purpose, mind you) of sexual reproduction is to stay one step ahead of illness and disease. The dull males have dull coats or unattractive tails because their immune systems have been pushed to the brink by bacteria and viruses. They're literally full of them. The immune system of the pretty males, however, is humming along. The little critters are knocking at the door, but they're not getting in. Females that mate with the dull males often have offspring that die young because their immune system is poor. The females that mate with the pretty males have little ones that thrive. This is how picky females and pretty males evolved, at least for birds.

It seems to be true for people too, but for us and some other species, there are more twists to this story. For most birds and people, both the males and the females watch over and feed their young. This means that the lady cardinals have to compete with one another—*female-female competition*—to get the pretty male to be their social partner, the one that will help them care for the little cardinals. They get the worm and eat it too—that is, they get the healthiest male as their social *and* sexual partner. At the same time, many other cardinals get stuck with the duller males. These ladies also want a piece of the worm, and this is where the EPCs, or extra pair copulations, come in. The behavior of these lady cardinals—and DNA fingerprinting to determine the little ones' dads proves it—indicates that they want the dull males around to help feed their young, but they want the pretty males to sire them. The dull males are not too happy about this situation. They keep an eye on their lady and take to the wing to keep her from dallying with the spiff and shiny males. Sometimes the dull male succeeds and sires the little ones, and sometimes the lady gets her way. She gets a healthy immune system for her offspring and a hard-working, if not a bit dull, male to feed and protect them. I'm not saying that the battle of the sexes simply comes down to keeping one step ahead of intestinal worms and other such nasty critters, but it is an important part of the story.

You're probably thinking, *Well fine, but does this really explain why and who people choose to mate with?* These are issues that evolutionary psychologists are beginning to tackle, and some early studies, as examples, indicate that good-looking and physically vigorous men appear to have better immune systems than others, but it's not a perfect relationship. People tend to overestimate the health of really good-looking men and women and underestimate the health of the duller types. For those of us in between, looks and vigor do appear to say something about our immune system. Darwin didn't know about the immune system, but he did suggest that women, like the lady cardinals, prefer spiff and shiny men—not men donned in bright red tuxes and hopping around in ash trees, but rather men who clean up nice. These men are healthy enough to cut a good dance and wealthy enough to sport the latest fashions. Sometimes they drive a shiny red sports car, the type that most other guys can't afford (a friend of mine leases one in an attempt to fool women).

These guys are flaunting their wealth, a type of symbolic phallic display. In other words, they got big ones. Like the successful male cardinals, most of these guys also have their own ash tree. All of this helps to attract the ladies. But this is not enough. The women not only want to move in, they want to keep the bedbugs out. They want to have their cake and eat it too. They want the healthy, vigorous men—those with a good immune system—who have a lot of stuff, not ash trees, but cash, cars, whatever it takes to raise healthy children successfully. It's not just good-looking guys with stuff, but guys who will stick around and share what they have. Deep down, women see guys as resource objects they can use to secure profitable traits for their children and that will stay around and help raise them.

The guys have some of the same preferences as the women, but there are differences too. When it comes to looks, guys are even pickier than women. Many women I know or have heard from, whether I liked it or not, have told me that guys' focus on attractiveness is superficial (even "stupid"). It turns out that the kinds of superficial things that guys find attractive and like to look at are correlated with fertility. Again, it's not a perfect relationship, but attractive women are more likely to give the guys babies. Of course, there is more to men than this. Many men in fact rate personality (compatible interests) above looks when it comes to picking a wife, but looks still count.

The punch line is that there is a rhyme and a reason for doing it, and for the battle of the sexes, that ensues once a species evolves to reproduce by mating. It all started as a way to deal with viruses, bacteria, worms, and all the other critters that want to do their business and make their living at our expense. The birds and the bees do it to outwit illness, and that's why we do it too—at least one of the reasons. However, we have progressed since the initial evolution of sex. Humans also do it for intimacy, to maintain relationships, and sometimes just to have some fun. People, in fact, are probably better than any other species (bonobos, a cousin of humans, may give us a run for the money) in thinking up new and better ways of doing it, but what does this bode for the evolutionary future of having sex? Maybe nothing. One of Darwin's most important insights was that the mechanisms of evolution are not forward thinking. We may evolve further, but from where we sit now, any such change may or may not be "progress."

Still, when it comes to why, where, and with whom we do it, culture and social mores matter, and progress can be made on this front. There are rules that say where and with whom you can do it. As pointed out by Mark Flinn and one of his colleagues, Bobbi Low, probably one of the most important of these rules is socially imposed monogamy: you go to jail if you marry more than one person at a time. In most cultures, rich men get many wives, and poor men go wanting. These poor guys do not always accept this situation and in fact can get rather testy, or downright violent, at times. When social rules put the brakes on the reproductive interests of wealthy guys—the rules for socially imposed monogamy developed over hundreds of years in Western culture—the result is that they can have only one wife at a time. This might be too bad for the rich, but it has important benefits for the rest of us. Most men get mates and eventually marry, and thus they are much less aggressive than would otherwise be the case. Women also benefit. They get the exclusive attention of their husband, and their children don't have to share him with another wife and family.

Understand that evolution is not really about progress; it's about adapting, across many generations, to ecological and social change. These changes may or may not be progress from our current perspective. Progress for us is more likely to come from the rules we develop for our behavior—rules that influence how evolved tendencies are expressed or not expressed. Many of these rules are about why, where, when, and with whom we mate. These rules can and do vary from one culture and historical period to the next. Maybe progress is just allowing different strokes for different folks.

The First Mantophasmatodeae

Sharon Carter

New insect order found in Southern Africa.
—*National Geographic*, March 2002

No species was created without a purpose.
—Shabbat 77b

A newly discovered stick insect has praying mantis
arms and an appetite for its family and friends.

Males are created from dust. The female
blames her parched exoskeleton on her mate

though it was she who wrangled a Rome apple
from the Arbor Vitae and offered him a slice,

forgetting he was carnivorous. Or so the story goes.
She was made from a rib and could at least claim substance.

There were too many alters to conflict and decay—
her offspring fought over who was to become the shepherd,

or agrarian. Cain ate Abel and Canaan crushed Jenin.
Hamas ambushed Jerusalem. Ramallah and Gaza prayed

to different deities, which led to much claw rattling.
The house of Abram was demolished in Hebron,

after which there was beheading among both family lines.
As it was in the beginning, is now, and ever shall be.

Warren Bennett, *Here Comes the Neighborhood,* 2002

Prismatic Progress

Dorion Sagan and Jessica H. Whiteside

"Is there any proof of God's existence?" Bill said.

After a pause, Tim said, "A number of arguments are given. Perhaps the best is the argument from biology, advanced for instance by Teilhard de Chardin. Evolution—the existence of evolution—seems to point to a designer. Also there is Morrison's argument that our planet shows a remarkable hospitality toward complex forms of life. The chance of this happening on a random basis is very small. I'm sorry." He shook his head. "I'm not feeling well. We'll discuss it some other time. I would say, however, in brief, that the teleological argument, is the strongest argument."

"Bill," I said, "the bishop is tired."

—Philip K. Dick, *The Transmigration of Timothy Archer*

1 *The Animal*

It must have been a dream. Scampering through the woods, we saw them—the lost animals. Each one stranger, smarter than what we have today, evolutionary oddities from another place and time—not necessarily the past, nor the present. There was a side-burned lamb child, and furry fleeters of several shapes and sizes—most were mammals, or mammal-like, although there could have been a reptile among them. These are the kinds of animals that lead you into the woods and put a spell on you, making you wonder if they ever existed outside the memory or mind. They could well have been familiars of Merlin Ambrosius, that child of a princess and a demon incubus, the magus who predicted the future and sent Stonehenge from Ireland to the Salisbury Plain. I doubt we will live to see the red or white dragon, as the vast majority of us have been born between the great dinosaurs and their rebirth as tiny genetically engineered pets.

But there is another world of animals, one that cannot be foreseen. It is from this "bestiary" that I wish to report. For I have seen—in a direction orthogonal to evolutionary time—: a female unicorn. No, she had no horn; yes, she is a woodland beast, sly as they come, and yes she is rarer than her mythologized brother—otherwise perhaps you would have heard of her. It was her I glimpsed—more striking than the bear-boar that came running toward me in my dream, forcing me awake. However fleeting, she is more real than the greatest idea of any philosopher.

2 *Lure of the Crystal*

In moments of maximum clarity
I imagine the most minute crystal
obscurely scaled, whale-boney.
I used to let my unconscious go to work
figuring out ways to encounter women.
Now I let it go to work
figuring out ways to strip the cover off the universe.

The Chinese say that there used to be ten suns. One day a drunken warrior shot them out of the sky. On a clear night, you can see the image of his rabbit on the moon.

3 *The Unmixing*

Like Nietzsche's last man, who, before the gruesome feast, burns his books amid the distant sound of his own voice, laughing atop the mountain, I forgot—we all forgot—that we wrote this hypertext or that we are now writing—here, now—the pages we will flip to later, making a Möbius stripmark, a bookplace in spacetime.

Historically, Platonic physicists, enamored of eternal equations and their entrée to the mind of God, don't believe in time, so progress for them is an illusion, and evolution is a nice trick—something you see from the inside as you try to piece together the cosmic story of which you are a part.

I met University of Tübingen mathematician Otto Rössler in Madrid. A nice man, old school, hunched over his overhead as he made the notes we all need to understand the ultimate nature of reality. Anaxagoras, that first evolutionist, he came up with ΧΑΟΣ. Chaos; made a worldview of it, before Einstein or Darwin, Newton or Galileo.

On my birth date, in my day book, I sat in front of Rössler and took notes, in a hunched and minuscule scrawl, about Pauli's primary chance (quantum indeterminacy) and how church father Gregorius picked up Anaxagoras's theory of perichoresis, that the world is a perfect mixture, and has already lasted an infinite amount of time. Nonetheless, Anaxagoras postulated, noös—mind or spirit—could theoretically unmix the mixture, reconstructing the pristine original heavenly cosmos from the present chaos. Father Gregorius applied the theory to the presence of wine in water, of Christ in the world of the father, the son, and the Holy Ghost when no one else (and certainly not Science) wanted it. The dogma was all but forgotten. But not now, no—for perichoresis, sharing the same Greek root as *peripheral,* or *perineum* for that matter, means unmixing. "Everything existed forever; the world was a perfect mixture," said Anaxagoras, and the word he used for mixture was probably ΧΑΟΣ.

Before James Clerk Maxwell's impossible demons were christened angels separating 1's and 0's on laptops, Anaxagoras fluted the news that ΧΑΟΣ could unmix the mixture. The question for Rössler, thinking algorithmically, was whether perfect recurrence, within our realm, is possible. He questioned mathematician Steve Shale, disputing the notion that the unmixing, perichoresis, was impossible even after infinite time.

Over beer and tapas, I put forth some questions to Rössler. His study was of endophysics. A kind of fractal situation, I was given to understand, with us the cosmic equivalent of lawful computer programs, and God—or rather the demiurge, outside—perhaps too busy to check on the creatures upon which he had bestowed free will. Free will came up because this modern chaos seemed just a tad too deterministic. But Rössler had high hopes; he was optimistic about the possibilities of determinist chaos, that it would unravel reality's dense tangle.

"I have a Cartesian solution," he admitted. Our conversation was overheard by a biologist.

"What is 'endophysics'?" she asked pointedly.

"It is the 'end of physics,'" he tittered.

At dinner I asked Rössler whether the unmixing was like the experiment theoretical physicist David Bohm mentions, the one with the viscous liquid that blends with ink when you stir it—only to separate out again, remaking the original droplet, when you reverse directions, and "unstir" it.

"Yes," he said.

I asked him if he believed in God.

"Yes," he said. "The idea of a heavenly father is a beautiful idea."

He then told me his idea, opening it with the warning that the great American physicist and drawer of naked women, Richard Feynman, had given up physics when a vision assailed him: his hometown, New York, ruined by nuclear bombs.

Oscar Wilde—who died in the same year as Nietzsche—said all good ideas are dangerous. This was Rössler's dangerous idea:

Rössler's idea—to cure animals of autism—was doable, he insisted. The theory was based on the idea that human infant's autism disappears once it buys into the mistaken notion that its mother's happiness is actually friendship.

"The machinery of love, if you will," said Rössler.

5 *There Is No Four*

Such is the nature of of.

Because we are in a Fibonacci series here—with pieces missing, as in the fossil record. Jumping Jack Flash is a gas, gas, gas: ΧΑΟΣ.

Ludwig Boltzmann, the father of modern thermodynamics, was a great admirer of Darwin. He applied statistical methods to inanimate nature, as Darwin had explained life's changes through the accumulation of numerous random variations, only some of which persisted. Boltzmann watched molecules and imagined atoms (whose existence was still scoffed at by his colleagues) bouncing around probabilistically.

Inanimate nature was also made of populations—not of living things but of unseen particles. And the particles assumed the most likely positions. Merlin nature shuffled, making unlike like, removing the imp from improbable; smoke rings gently blew from the lips of sweetheart chance, never to make an earthly return.

However, all were not convinced. French polymath Henri Poincaré showed that, in infinite time, everything that can happen will happen, and not once, but an infinite number of times. You will be reading this line. And you will be reading it not once, but an infinite number of times. Now will explode; the two of clocks will become the queen of forever.

Thus there was a fly in the ointment of Boltzmann's derivation of linear time from reversible mechanics; the lubricant was also a glue. Later, Einstein and Gödel would try to improve upon Boltzmann's derivation of linear time from equations governing the reversible actions of particles. But Einstein gave up, inventing relativity theory instead. It was easier. Boltzmann, a lifelong migraine sufferer, killed himself.

Nietzsche took the bull by the horns. The übermenschen or "superman" wasn't a Nazi but the man who could look squarely into the eyes of eternal recurrence, the "heaviest thought"—everything that will happen has happened, and not once, but an infinite number of times. You choose to do it again. With joy and, as Heidegger put it, "active forgetting." It is why the last man in Nietzsche's *Zarathustra* burns his books with laughter. Here is where we get so close to the truth that it kicks us in the teeth. The "almost magic" of turning the page.

8 *Solar Archon*

Yes and no.

Librarian, student of the Middle Ages, and proto-deconstructionist Georges Bataille (the half-mad author of *L'histoire de l'oeuil* who André Breton kicked out of the surrealists because some juxtapositions were just too much) believed himself the veritable reincarnation of Nietzsche. Deeply familiar with the power of the universal onto the personal, the powerful notion of the individual as microcosm of the macrocosm, Bataille did to thought what the French Revolution did to politics: cut off its head. Apart from his explicitly headless writings, Bataille achieved his modern apostasies by linking what had not been linked, and unlinking what had. The highest point he argued in "The Solar Anus"—whose very title mixed high and low—could be reversed and made most low. He spoke of parody, and of crime, and linked the anciently evolved sex act to country-banging trains. His madness, his sexualization of nature, and his perverse attempts to transgress politics (split, at the time in France, into fascism and communism) led him to one of the greatest parodies of progress ever sprung forth by the mind of man: Bataille's robust rendering of evolution.

In Bataille's vision, human evolution progresses upward, like an erection of Earth's surface, into the upright animal we know as man. Only the act is not complete. As Bataille points out, for the act of evolution to be complete—for evolutionary progress to reach its pinnacle—the placement of the eyes should not be arrested at an angle parallel to the ground. No, for evolutionary progress to reach the mad gnosticism of its erotic potential, the eyes would have to migrate to the head, where they might best enter congress with the sun, that "most abstract object in the universe" (because it is always there, but you cannot look directly at it without going blind). Real evolution, Bataille silently screams, in his parody of a parody which sacrifices humor upon the altar of the sane, would entail the explosive ejaculation of the entire contents of the human person toward the Sun. Mother and child reunion—or father and child, depending on your language and how you like your solar caress.

Here we have a joke on progress, a joke which is not a joke because it contains the sense—which the deconstructionists under Derrida were to run with—that nothing makes sense except in the context of an All, which is denied to us by our emplacement within it as mere parts. Engorgement precedes understanding.

13 Technomystical Excursus

Here is the rub, and it may be bad luck. We are expanding. I have argued elsewhere, under my own name, that we are in the midst of a fourth and perhaps final Copernican deconstruction—that of life as an energy process linked to other nonliving processes that assemble complexity from the potential of their already complex surroundings. The third CopDecon, the stage that precedes us, was begun by Wohler's synthesis of urea from ammonium cyanate in 1828, which showed living things to be made of the same substances as nonliving things. Here began the cosmic awareness that we are, as Dad put it, "star-stuff" or, better, starsh*t. Yes, spice boys and puppy dog tail girls, the oxygen in the air and in your blood, in your brain, helping you to understand this sentence comes—along with carbon, neon, and silicon—from the gamma ray and neutrino bombarded outer layers of a neutron star, the exploded remnants of a type II supernova. The amino acids once known only in life have now been synthesized in laboratories and retrieved from extraterrestrial meteorites. The second CopDecon (we are working backward now) was inaugurated by Darwin's 1859 colossus, which removed humanity from its special place among creatures even as Copernicus and Galileo had removed the still Earth from its place at center stage.

Exit stage left: the fourth CopDecon is the realization that humans, the most revered and intelligent sentient beings, are but one of many open systems tending toward the energy reserves that supports their cyclical organization; if life could do what it wanted, it would turn the sun into itself. Like the whirling tornado, we whirl too, but our cycling is not just a matter of the emergence of an effective system of whipping winds: our cycles are biochemical and transgenerational. Instead of reducing barometric pressure and disappearing as a hurricane does, we help reduce the solar electromagnetic high-to-low quality quantum energy gradient between the 5800 Kelvin (pretty hot) Sun and the residual background radiation of 2.7 K (nippy) space. Once we have finished with the Sun, or it has finished with us (becoming a red giant and boiling up what's left of the oceans), we too will disappear. It would be nice to get to other planets around other stars, but it may not be feasible. The universe is expanding and, as we sit here twiddling our technological thumbs, the other stars continue to fly away from us, at a very fast rate.

The perfect sentence doesn't describe this.

What is the meaning of this?

We are being abandoned by the stars themselves, and any scientific reckoning would, of course, explain it as naturally derived from the physical laws of the universe. But one of these laws (curiously, at once the most general, and the most "imperfect" one—because, as Steven Hawking points out, it need not always apply) is the second law of thermodynamics. The second law, put forth as the universal tendency for entropy to increase, can be restated to say that "nature abhors a gradient." Not knowing this law and being alive is to think you know all about love without ever having read *Romeo and Juliet*. Spring comes, snow melts, and Shakespeare's human staff goes flying toward the heart of longing; his actors go flying about in the human equivalent of Brownian motion. So we may be stuck here, aimlessly, going nowhere, until we die. Or we may be part of the unfolding by which the All grows and comes to realize—again—what It has done.

What if everything and its opposite were true, and this was the only possible universe in which such a situation could be?

That most brilliant if horrifying proof of being alive—the one purveyed in the Hindu Vedic literature, and later by the gnostics and their spiritually trapped offspring—is gripping in the extreme. If they are right, the physical realm is an illusion, and its mental apotheosis would be universal implosion, recreating the stellar unity science detects in its measurements of red shifts and the background microwave echo of the big bang. In this way, the Kali Yuga would end, and we could all wake up from this bad dream of time and incarnation; we could end our tenure as gas-emitting chunks of flesh, skin-encapsulated egos. We would wake up, and we would no longer have to read essays like this, or poems about lost animals. Because the lost animals would be found, and then some. Thus would reform the One.

In this way we see that the Earth is an egg, and that it is developing.

Three things can happen to this egg. It can be boiled. It can break.

Or it can hatch.

But the gnostics may also be wrong. The cosmic spring that returns us to the primordial energy of the big bang through the looking glass of black holes may no more be in our stars—stars which are to be found inside us—than the end of time. Anaxagoras may be right. The universe may be a perfect, infinite mixture, unmixable by mind: universal death may not be in the cards. I write red heart red heart red heart. But the slot machine comes up black spades.

Buddha had three imponderables. (1) Don't try to understand the mind of a Buddha. Only another Buddha can understand the mind of a Buddha. (2) Don't try to understand the laws of karma; they are too complex. (3) Don't think too much about the beginning of existence—you will go crazy.

Don't think too much about the beginning of existence—probably because existence didn't begin. It already was. Is. And will be.

Such is the nature of progress.

II

Steps from the Cave

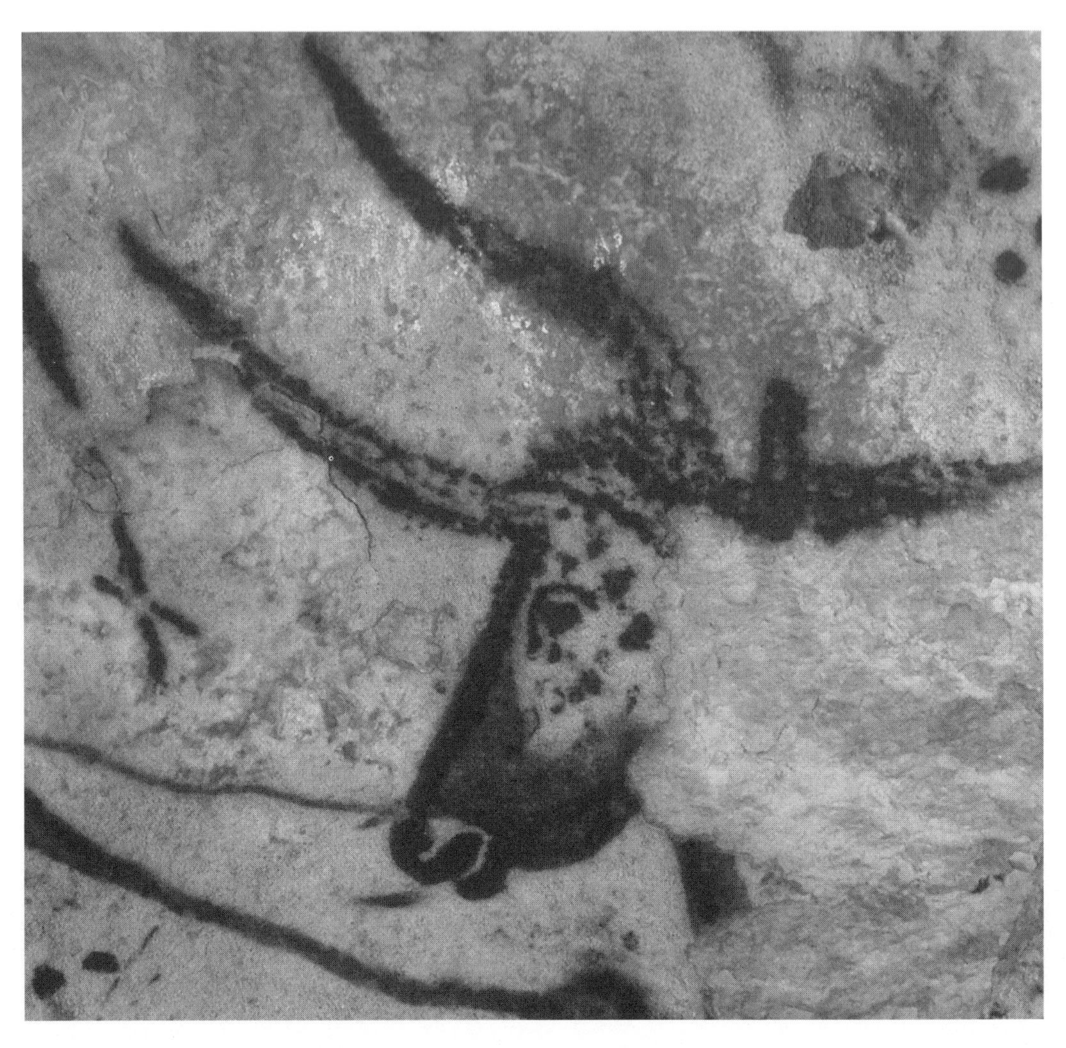

Norbert Aujoulat—CNP—Ministère de la Culture. Head of large bull at Lascaux

Of Caves and Humans

Ellen Dissanayake

As a child, I loved the idea of caves. More sensible children might have imagined them as inherently fearsome, concealing wild animals, madmen, or outlaws. Instead, I dreamed of caves as enchanted places for exploration, far from adults and the ordinary world, with hidden passageways and underground rivers. Or, as in a Beatrix Potter story, one could live in cottage-like rooms, their curved walls and ceilings made of earth instead of plaster, with little ledges for toys and perhaps a few conveniently placed stalactites and stalagmites for furniture—a fantasy house.

Visits to real caves during our family's summer vacations did nothing to alter my sense that they were places for play and exploration. The two or three caverns we visited were well lit, easy to walk about in, and jazzed up here and there with recorded music and colored lights. We saw blind fish in pools of water. We heard drips and felt a chill. Such details only added to the cave's prospect of romance and adventure.

Long after my own children were grown and I had begun writing about the human impetus to make and experience the arts, I made a pilgrimage to visit some of the most famous caves of all, those in southwestern and south-central France that contain relics of Paleolithic humans. Contrary to the inaccurate stereotyped notions about "cave men" in popular cartoons and greeting cards, our ancestors used caves not to live in but for sophisticated activities that we now classify as "art"—painting, engraving, drawing, carving various degrees of relief sculpture on rock walls, and in some cases molding clay figures. Some of these images are among the best known pictures in the world, as famous as the *Mona Lisa*. These painted horses, bison, and other large animals are usually the first pictures in art history books, typically in a chapter titled "The Beginning of Art."

My desire to see the Paleolithic caves had little if any residue of the naive infatuation with "caveness" that I had felt as a child. My purpose was to experience the art they contained, as I would go to the Louvre to see the *Winged Victory* or St. Peter's Cathedral to see the Sistine Chapel. Additionally, having claimed in my writings that artful behavior had been an intrinsic part of human nature for many thousands of years, I was eager to measure my own response to some of our forebears' most esteemed visual images. I expected to be moved by the paintings, but it was a distinct surprise to discover that I responded to the "architecture" of the caves as well. Indeed, I found that my reactions to the art were inseparable from their setting, which was decidedly not the cozy hideaway of childhood daydreams but, on the contrary, a profoundly alien and inhuman place.

The fact that our ancestors chose such places for their images (and for the rites that are thought to have accompanied the images) corroborated for me an insight about artmaking that I have developed in my work. Artists everywhere make the ordinary *extra*-ordinary or special. Although the statement sounds simple, that is what artists do, whether today or thousands of years ago, and whether their medium is paint, clay, words, sounds, movements, ideas, or feelings. Although the Paleolithic images are "inside" caves, just as more recent paintings and sculptures are inside museums and cathedrals, this placement required that the artists overcome a natural animal reluctance to enter such a dark (and hence possibly dangerous) and humanly inhospitable place. That they did so, often navigating perilous descents and wriggling through narrow tunnels, indicates that they deliberately chose to humanize and find sacredness in unearthly sites that were unlike anything else they may have experienced. In other words, our prehistoric ancestors not only created their extraordinary images but were moved to create them in perhaps the most extraordinary place of all.

My revised adult view of caves and humans conflicts not only with my childhood imaginings, but with the popular opinion expressed by Paul G. Bahn in his hugely informative, sensibly argued, and provocative book on Paleolithic art, *Journey Through the Ice Age*. Bahn claims that although our ancestors did not typically live in caves (as opposed to cave mouths and rock shelters), the footprints, knee prints, and handprints of children found in some of them indicate that children "were clearly not afraid to explore the far depths, narrow passages and tiny chambers of caverns, whether alone . . . or with adults."[1] He suggests that the finger holes and heel marks at Tuc d'Audoubert were made by children playing while adults fashioned the clay bison.

After visiting the caves myself, I find it difficult to believe that any living creature, except one like a bat or bear, an animal that has evolved to live in caves, would want to spend any time deep inside them. To an earth-adapted creature, a cave is a markedly alien environment. To deliberately place one's pictorial efforts in that setting is a choice that is as significant as deciding whether to depict a bison or a horse or whether to use red or black pigment. Although my visit to the French caves altered my naive childhood fantasies about caveness, it expanded my insights into the mind of artists and enlarged my understanding of the similarities between Paleolithic people and myself.

Gargas was the first site my French friend and I visited, for the simple reason that she lived nearby. Although as we entered I had little idea of what lay ahead, this cave was to become the prototype for my new appreciation of the uncanny otherness of the world inside the earth. Gargas is one of the many cave sites in France that is on private land and managed by its owners rather than administered by the State. Consequently, the times of access, interior amenities (such as lighting, walkways, competence of guides, and extent of their route), and even availability of postcards depend on the conscientiousness, the means, and the goodwill of the owner.

At Gargas, our tour language was French, and a group of perhaps ten followed in dim light behind the guide, a young man who from time to time stopped and shone a flashlight on a wall, outlining the engravings of animals with his penlight. Unless one was at the front of the group, standing at the proper line of sight, these marks had to be taken on faith.

It was not the depictions of animals that was the most remarkable human activity in Gargas: it was the handprints—approximately 150, which dated from as early as 28,000 years ago. The black and red negative handprints (pigment has been blown or applied around a hand that is pressed to the wall) occurred in compact groups, as if made on one occasion or on many occasions in reference to one another, and most had one to three fingers missing at one or even two joints. The guide explained that no one knows why, or even whether, hands were mutilated. Perhaps these missing fingers were folded down when the paint was applied. Did the palms of the hands press the wall or face outward? With negative prints, no one knows. But because handprints are such a natural human sign—appearing wherever there is wet cement and on nursery school masterpieces—one instantly feels a sort of recognition: *These prints were made by someone physically and psychologically like myself,* followed by a stab of surprise or even shock when one notices the stubs of fingers. Why would someone come here to press a mutilated, or intact, hand to the wall? Anyone can guess, but no one will ever know.

The mutilated handprints, however, only underscored the strangeness of the cave itself. Gargas was far removed from the warmly illuminated and cozy hideouts of childhood fantasies; it was vast and endless and dark—its perimeter lost in faraway dimness. Although we could not see it, we were told that the ceiling was as high as that of a cathedral and the width of the main chamber similarly wide, some thirty to forty meters. The owner of the cave had chosen to provide only the minimum amount of light and few, if any, wooden walkways, so we found ourselves stumbling on uneven ground, stepping into puddles, and occasionally nearly bumping into pillars of stalagmite.

I realized that the familiar hills and cliffs that we see every day give no more idea of what lies inside them than our own familiar bodies. Before anatomy textbooks existed, who would have guessed that there are lungs, a cage of ribs, a heart, a stomach, and an intestine? Who could have guessed what these actually look like? But at least these bodily organs and structures, having evolved for the purposes of life, have a comprehensible symmetry or regularity of form and proportion. What characterizes most caves is *incomprehensibility.* Their space is chaotic—undesigned and unarranged. Or if "arranged," not by humanly comprehensible forces but by such things as ancient shifts and shrugs of earth, diverted underground rivers, hidden whirlpools, all rearranging mindlessly over thousands of years in utter darkness. There is no symmetry, no pattern, and thus, at least for me, no human welcome.

Gargas introduced me to the utter inhospitality of caves, which I was to experience in all the caves we visited, although to a lesser degree at Lascaux, Pech Merle, and Cougnac. Had we begun our cave visits with Pech Merle, for instance, my overall expectations would have been quite different since its presentation and atmosphere were almost the polar opposite of Gargas. Managed by the State and a conscientious curator, Pech Merle (discovered in 1922) was extremely "visitor friendly," with a pleasant waiting room at the entrance and a little museum in which we viewed an informative film and looked at Paleolithic artifacts in vitrines. We followed our well-trained guide along carefully lit fabricated walkways and steps that made getting around as easy as walking through a hillside village.

In contrast to Gargas, the good and occasionally dramatic lighting at Pech Merle allowed us to see the varied geological or "architectural" features that were reminiscent of a Disney set for a cave except, if anything, the real cave was more varied and outlandish than any human imagination could devise. It was like being inside a monster's body with weird fluted or lobed or twisted or smooth or convoluted or granulated organs and tumors on every side. Some formations were brown, others buff, some like enlarged versions of crusty French loaves. One could marvel at nature's handiwork—unpredictable and strange but somehow not disquieting. Whereas Gargas had felt cold and inhuman, Pech Merle seemed almost organic, its features something like coral or fossilized remains. Perhaps the lighting helped us forget the drips and puddles, the chill, and the hard mineral-encrusted surfaces that we would surely have been more aware of with less ingenious illumination.

But eventually I would have understood that most of the caves, like Gargas, have an unsettling atmosphere, even when it is disguised and temporarily forgotten, as at Pech Merle. Caves are "alive," but in a sinister, proliferating, mineral way with cold water oozing over surfaces or dripping, and glittery sliminess here and there. The chill begins at one's outer clothing and gradually reaches one's bones. The difficulty of seeing well or seeing much is unnerving, and all the recesses and great heights suggest mystery—even menace—just beyond visual and tactile reach or effectiveness.

Because caveness is so distinctly alien, it struck me quite forcefully just how exceptional it was for Paleolithic humans to have begun to enter, explore, and make use of caves, as at Niaux—where the Salon Noir, with its extremely large paintings of twenty-five bison, sixteen horses, and ibexes, is some 800 yards from the entrance and has a domed roof some 250 feet above the floor. To visit this giant room, each pair of our party of twenty shared a hand-held lamp as we tramped up and down through narrow and wide passages, ducked under low ceilings, and traversed many spacious rooms.

Like Gargas, Niaux had its drama and mystery, but there the entire experience became overpowering as well, too much to take in—too large, too far, too deep. I could not imagine what sort of events took place there. To my surprise and disappointment, I found that when we finally reached them, the black and red paintings were more impressive and beautiful—more *visible*—in books than before my eyes. Perhaps having seen the images isolated in books, I found them too familiar to be seen with fresh eyes, even in situ. Although their placement that required winding through narrow passages into deep and inhospitable interiors was impressive to think about, the images themselves were less affecting than I had anticipated.

I noticed that after leaving each cave, I responded to the French countryside outside with ardent pleasure, as if I were returning home. In its terrestriality, of course, it was instantly familiar, like a loved one's face after an unwanted separation. Smells of earth, songs of bird, the colors of the sky and trees and other plants, the currents of warm living air on my face were so obviously and indisputably agreeable to life. With my feet on (rather than inside) the earth, the living landscape became a source of safe, imaginative reverie. I loved the rocky, wild cliffs and idly wondered whether their camouflaged steep slopes concealed still other unknown decorated caves, sealed and secret for millennia.

Le Rouffignac was a different kind of cave, notable for being "dead." That is, there was no mineral activity and wet slipperiness, no stalactites and stalagmites. It was something like Alice's rabbit hole, although horizontal rather than vertical, and if miniaturized, it could have been the burrow where Thumbelina spent a winter with the kind rat. Unkind cave bears apparently once hibernated or hid there, and we were shown their impressive ancient claw marks high on the walls. Resistance fighters hid there during the Second World War, and local people had been leaving graffiti on the walls over several hundred years. We rode on a little open train as on a carnival ride through tunnels called "galleries," which we learned extended over eight kilometers. The major feature was a far room with a painted ceiling covered with black-outlined mammoths, bison, goat, and ibexes. Many of the drawings were skilled, with overlapping legs, a recognizable haunch, and male sexual organs. It was explained that the floor of the room was originally only a few feet below the ceiling. The painters had to lie on their backs like Michelangelo. Now we could stand and look up at the figures and could even have touched them if that were allowed. Ironically, I found that while the earthiness of Rouffignac was not alienating in the way that "live" mineralizing caves had been, it was nevertheless not a place where I wanted to spend much time. Its dry dustiness was unwelcoming, like being in an underground warren, not a place of comfort, much less of sanctity, romance, or mystery.

The excellent lighting and convenient walkways at Pech Merle (whose visitor-friendly atmosphere contrasted with the inhospitableness of Gargas) made it possible for its images and markings to be as deeply moving to me as any I've seen elsewhere. In Pech Merle's Grand Hall is the famous dotted horses panel, with two spotted horses, facing away from each other, on a background embellished by dots and hand stencils. I wanted to study this panel as I would any other fine and meaningful painting: Why are the heads so small? Why are their hindquarters overlapped? What are the dots for? The hands? What is the difference between black and red? Look at the wonderful lines of the horses and their relationship to the shape of the rock! There were images of a female bison, a wounded man, the Great "Black Frieze" fresco (a bull with his breath showing), and more red dots. In the "Diverticule of Femmes" (Passage of Women) were red handprints and dots, reminiscent of blood, tantalizing with imagined but unknown significance.

There were many other fascinating marks in the various rooms, including red and black dots, doubled dots, handprints, quadrangles, and tectiforms, many in association with images of animals. All these signs reinforced the impression that it was people (and artists) like ourselves who had been here and consecrated the space for their own purposes, their own meanings. As at sacred sites from more recent historic times, the placement of these prehistoric paintings did not seem random. One sensed not only an artist's mind but an organized mind: the images and signs had a considered place.

These impressions were only increased at Lascaux (discovered in 1940), whose renowned paintings are from approximately 17,000 years ago, around the same time as those of Pech Merle. Today it is as impossible to see Lascaux with unprepared eyes and mind as it would be to visit the Taj Mahal or the Grand Canyon. Images of the place from books and postcards are easy to see, and in fact hard to avoid. Whether the actual experience lives up to the reality, such sites offer the possibility of a revelatory experience.

To protect the paintings from the organic world's degradations brought in on the breath and clothing of visitors, the French government has restricted visitors to everyone except archaeologists and other people with professional interest in the paintings. For the ordinary tourist, a *faux* Lascaux, accurate in every detail, has been built above ground near the actual site. I was one of the blessed who received permission to see the original Lascaux, because I had been brazen enough to write four months in advance to the *conservateur* in Bordeaux, enclosing promotional material about my books that I hoped would establish my bona fides. Some weeks later, he replied and bestowed an assigned date and time to appear.

The entrance to this marvel of human accomplishment looks like nothing more than a steel garage door, painted green as I recall, down an ordinary sort of driveway cut into the ground. This unprepossessing entrance, which could just as likely have led to a storage shed, was visible only after our guide appeared and unlocked the unmarked gate where we had been instructed to wait. Inside the garage door was a small room that served as a sort of airlock, where we stood until the outer doors were closed. Another door was opened, and we went down some metal stairs into a larger anteroom, with rock walls. There was a footbath of disinfectant into which we dipped each shoe sole. (At many sacred shrines around the world, visitors remove their shoes or wash their feet. The purpose of this ritual at Lascaux is more pragmatic than honorific: to remove as much bacteria or fungi from the outside world as possible. Yet this respectful gesture seemed more than appropriate, and we were glad to comply.)

After such routine precautions, it is a considerable surprise to find that the next door opens right into the great rotunda—the Salle des Taureaux, the great destination itself. Visitors do not have to trek a kilometer or two into the bowels of the earth but rather simply descend into a large room. And there they are—a surrounding swirl of enormous animal shapes on upper walls and ceiling, almost within touching distance, no farther away than the stained glass windows or wall paintings in many European churches. Compared to any other known sites in this region of France, the animals are much larger, more polychrome, and have more drama and vitality individually and as a composition.

This monumental frieze depicts various large, powerful animals, some five yards in length. They seem to race and leap, with immortal energy. There are five large bovids, two of them facing the other three. Seven horses to the left of these face in one direction. A chimerical animal has the hindquarters of a bison, the belly of a pregnant mare, the front legs of a feline, a mottled hide, and two long straight horns.

Both the paintings and their placement are spectacular. Even today, it would not be easy to make such large paintings, and it would be even more difficult with torch or lamplight and with primitive materials (blowing pigment or applying it with a frayed stick). The painting surface was not so close to the artists as at Rouffignac or some areas at Pech Merle, where we were told the artists (and viewers) had to slide on their backs through tunnels. Still, it is a marvel that the proportions of the animals are so convincing and the lines so certain. The guide showed us sockets for scaffolds that the painters had used, and said that stones hollowed out for oil lamps or for mixing pigment were lying about when the cave was first discovered.

The colors of Lascaux are lifelike and intrinsically beautiful—earth colors that are, at the same time, the color of animals' coats. Their application, with shading or two different colors side by side, and the painter's use of protrusions and declivities in the rock wall that give form to a shoulder or haunch, result in great realism and power. The animals drift across the ceiling and along the walls in a theatrically considered way, as if designed and even painted by the same hand. Unlike the other caves, there is little, if anything, that is tentative, arbitrary, or unskilled at Lascaux. The paintings are as effective as the iconography of any impressive architectural setting, a palace or church, planned as a breathtaking totality.

Other parts of the cavern work their own power and strangeness on the imagination. We were allotted forty-five minutes to see only a few of the various rooms and corridors—the falling horse tumbling around the side of a huge rock, the "swimming" deer, the deer with nine-point antlers, and the fat-stomached "Chinese" horses. We were not taken to the interior rooms with the famous falling sorcerer, the bison and his entrails, and the rhinoceros.

Unlike some other caves, where the great animals are engraved on dark, dimly lit rock walls and loom out of shadow, the light-colored walls and careful artificial illumination at Lascaux allow visitors to view the paintings as in an art gallery. As at Pech Merle, the natural translucent mineral depositions on the walls of Lascaux preserved the paintings as if with a coat of varnish. Although these caves were still "alive," with ongoing stalactite and stalagmite formation, they did not seem coldly inhospitable like Gargas and Niaux. Humans like ourselves had humanized them and turned them into sanctuaries, which today's French Ministry of Culture had further preserved and shown to their best effect, so that their inherent "caveness" became secondary and almost irrelevant to the experience of the paintings.

To my mind, the cave at Cougnac occupies a place midway between the emotional and physical discomfiture of Gargas and the emotional and physical accessibility of Pech Merle and Lascaux, and perhaps it was this combination of the mysteriousness of caveness and the intimation of humanness that provoked the unexpected richness of my response. Cougnac, discovered in 1949, is much less popularly known than Pech Merle and Lascaux, although its animal images are from the same general archaeological period—some 15,000 to 17,000 years before the present. Of all the caves we visited, it had the most mineralogical activity, being a forest of stalactites and stalagmites and layered, coated walls.

Our guide was a local man who remembered when the caves were first opened, and he obviously had absorbed knowledge of the place from boyhood. He asked whether we wanted the "geological" or the "artistic" tour, and we asked for a little of both. Walking backward as he talked, by some sixth sense miraculously stepping left or right just in time to avoid crashing into a column of stalagmite or avert a concussion against a looming stalactite, he pointed out the unusually dense forest of these mineral forms on the floor and ceiling surrounding our pathway. They were remarkably fragile, he said, too delicate even to touch, and he pointed out many that had been broken off by the Paleolithic people who had walked along the same route as ours. (They had since acquired little caps of new dripped deposits over the intervening 15,000 years.) It was difficult to pay attention when distracted by his virtuoso backward stroll, but eventually we stopped in what seemed to be a sort of circular room.

The guide directed our eyes upward toward a rainshower of thin, pointed white stalactites descending from the ceiling, so narrow and small and closely packed that one could not imagine space to insert even one more. These crystalline needle forms glittered like icicles, a spectacular frozen downpour. They reminded me of the Dilwara Temple at Mount Abu in India, which is made entirely—walls, columns, and ceiling—of intricately carved ivory that seems to proliferate like white moss on every surface.

The guide vocalized so that we could hear the fine acoustics; he surmised that there would almost surely have been singing or chanting here. He told us that he would turn out the light for a moment, and we were to turn around and wait. We did so, and were reminded once again how unnatural it is to be alive and warm-blooded in the middle of the earth. At Lascaux, Pech Merle, and Cougnac, we could temporarily forget where we were, rather like forgetting that you are in the middle of the sky when riding in an airplane.

When the light returned, we found ourselves facing a red-outlined ibex—almost a sketch—whose shaggy coat was given texture by mineralized pleats in the wall surface, as its haunch was enhanced by a slight bulge in the wall. Near it were emphatic mysterious marks. It seemed likely that the Paleolithic people who had made the image (and others that had been pointed out to us as we traversed the cave to this place) had intended that others stand in darkness as we did before the animal was dramatically revealed. To anyone being initiated into the group's sacred lore, the revelation afforded at this exact spot would be powerful and mysterious, and at least one archaeologist has commented that flickering lamplight makes the Cougnac animals seem to move. I could believe that this was the intended viewing spot because the animal, which was not large, was much less impressive when viewed from another angle.

As I responded so strongly to the unearthly beauty and strangeness of the site, I began to think that those who had made the image must surely have felt it too. Fifteen thousand years ago, people like me had seen in the round crystalline room a sanctuary—a little chapel or temple. Rather than being gargantuan as at Niaux or Gargas, an endless dark-walled labyrinth as at other caves we had visited, or a dusty burrow like Le Rouffignac, it was in human scale. Although it was not nearly so imposing and theatrical as Pech Merle or Lascaux, or the animal images so impressive or powerfully arranged, this particular glittery white room had an unmistakably special—call it spiritual or otherworldly—quality. I thought of how I had felt years before when I first saw the remote white Himalayan peaks impossibly high against an unearthly pale blue sky. The pure blue and white were unlike anything else in nature, especially the green and brown earth of the foothills on which mortals toiled and suffered. To human minds, they naturally seemed the abode of the gods.

Later it occurred to me that Cougnac was a place where *only* a human would find meaning or would assign meaning to its strangeness. Take any other animal in there—a dog, a cat, even a chimpanzee—and it would react as I did in the bowels of Gargas and Niaux or the desert of Le Rouffignac, wanting only to return to the light of day, the trees moving in the wind, birds singing, green foliage, brown earth with its organic smells, the familiar everyday world of nature, warmth, and life that we terrestrial mammals were evolved to live and prosper in; not the supernatural, cool, silent, dark, otherworldly world in there. No other animal but ourselves would recognize the Cougnac "chapel" or the remote Himalayas as beautiful or special, and therefore make them the home for their gods, leaving their marks on the cave walls as evidence of this recognition.

Among animals, humans are unique not only in recognizing the extraordinary in nature (what Kant called The Sublime) and responding to it with wonder, but in conferring specialness on what is ordinary, thereby surpassing or transforming it. A similar impulse, I believe, may have driven our Paleolithic ancestors to penetrate two kilometers into the farthest depths of Niaux and to squeeze through narrow tunnels to make or view engravings as at Pech Merle. This need or tendency to make ordinary things extraordinary, typical of all art, has characterized our species for at least 32,000 years.

At Cougnac, one feels astonishment at a wonder of nature, but unlike the large deep caves, or the Grand Canyon, or Niagara Falls, it has been humanized—is capable of being humanized—by the humans who used it. Responding to its strange, visionary beauty, I discovered that our faraway ancestors must have felt similarly and then were moved to add to its already extraordinariness with their own marks and ceremonies. The connection across the millennia that I felt at

Cougnac came not only from marveling at such considered and astonishing artistic creations but perhaps even more from realizing a shared humanness that finds inspiration, awe, and spiritual meaning in the great features of the natural world, whether remote white peaks high above the mortal earth or crystalline chapels hidden deep inside it.

Acknowledgments

I am grateful to Michelle Hankins, who shared with me the visit to the caves described here, and to Haystack Mountain School of Crafts, Deer Isle, Maine, where I began this essay during my tenure as a visiting writer in summer 2000.

Note

1. Paul G. Bahn and Jean Vertut, *Journey Through the Ice Age* (Berkeley: University of California Press, 1997), 10.

Lascaux. Pech Merle. Chauvet.

Andrew Schelling

Lascaux. Pech Merle. Chauvet.
Horse and bison murals sweeping in charcoal.
Moon antler carvings Siberia
Towers of handcut
trance induced
love pleasure temples Khajuraho.
Hovenweep calendar spiral high on a boulder.
Shila-na-gig cunt shrines.
Tun Huang's glittering Tara cliffs.
A young girl
folded with fox teeth
in Tennessee.

What bronze Pentagon skygods
put a stop to all that
nonsense?

Tyger Tyger

Andrew Schelling

The small outfit of contemporary techno-wizards who've taken up digs at The House of William Blake had some Americans by for a visit in May. Seventeen South Molton Street—last standing residence where William and Catherine Blake lived. The neighborhood's pricey these days, the global economy whirrs past its door, and sharply outfitted women go hunting for clothes. *Milton* and *Jerusalem* got written in second floor rooms, a sizable hand press stood by the window that fronts the street.

The good people in residence at the Blake house have almost no relics, but they did bring out a single archival box—stencils for *Milton*—also showed us wide modern drafting tables and high-end computer monitors used like bellows and anvil for angelic techno-designs out of Hell. Biscuits sat by the firebox where Blake once burnt coal. There was wine and a white British cheese. Warmed by the hospitality of our gentle hosts, and considering that the tiger Blake observed at the London Zoo had been brought out of India, I reframed a stanza of "Tyger" to Sanskirt. Had nobody tried it before? Surely some ganja-headed pundit of old Bengali renaissance days—?

> *shardula shardula ratrivaneshu*
>   *tejah prajvalan*
> *ko 'mrtah hasto va chakshu va*
>   *te bhimam rupam kartum shaknoti*

Early June, home to the Colorado foothills, west by a tiger's hair of the 105th Meridian. Icy mist holds the Front Range. It crawls down from the summits through boulder-choked canyons, leaving needles of frost on dark Douglas fir. Evening it vanishes upwards. Red Dakota hogbacks slip forth, a glimpse of smoky forest ravines that drop from Indian Peaks. Then precipice moon.

Who wandered these forests when Blake was setting Tyger to verse? Ute Indians mostly. A few tough Frenchmen out trapping beaver. And did he smile his work to see?

The region's dominant cat is *Felis concolor*—cougar, catamount, puma—mountain lion or painter—depends where in N. American space you picked up your speech. A Tupi Indian word passed to French trappers. Or archaic Greek, bent to the way things get said Upper South.

painter : panther : *Panthera tigris*

Caught in a coyote snare
on the Uncompahgre plateau,
I saw you there
thy tawny pelt
thy pelt philosophic & tattered
thy stiff drying deer-color'd pelt—

Blake died in 'twenty-five. Five years earlier, Dr. Edwin James went up Blue Cloud Summit, botanizing the tundra, and named the mountain for Zebulon Pike. In '06 Pike had gone through and put cat tracks into his army report. By which time Blake had turned upside down a full notebook, and was drafting Tyger and London on the same empty page. The Southern Rockies were still Louisiana—blank on maps in the London cartographer shops.

Hail catamount,
tawny end-of-tail flicker once glimpsed as the
mesa grass stirr'd
or felt dread feet when the stars
threw down their spears over high twilit
meadow alone—?

A scrape of dirt & debris, whiff of sharp urine
muddy track in the gloam.
Lay it down Tyger Tyger for humans—
& frame old symmetries
          new poems.

10: vi: 98

Rock Is Naturalist Scripture

Andrew Schelling

Rock is naturalist scripture. The deeper you go the older the story.
Pikas & squirrels scamper over the top, then spiral descent
from gone tooth & twig. Petrified bone sediment myth.
Or psychic fossil? Horsetail & algae glow green again,
come to life in car engines. Fantastic shapes, old as forests.
And now the likelihood we have in the world
as many diverse minds . . . "as there are
organisms capable of perception."

Evolution's basic
job—turning rock
to green growth.

Judy Johnson-Williams, *Zone of Turbulence* (detail), 2001

The Man Who Spoke to Stones

Stephen Miles Uzzo

The Secret Life of Stone: Enhancement

The sky quickly turns from pale to blue as the sun vaults over the New England heavens. Late August air invites the sun to shower the Earth with light. This small planet replies, sending back colors from the rich soil, the forest canopy, the river's moving waters. They fade only in the occasional plumes of dust, which rise from the powwow grounds as cars fill the parking field. "Powwow dust is good dust," I think as I walk toward a line of dilapidated trailers along the wood's edge near the Connecticut River. My eyes fix on an elderly man holding something fluorescent yellow, flopping back and forth in his hands. He sees me and waves. As I approach, I recognize him as LaVan Martineau. Meeting with him is my reason for attending the Connecticut River powwow. I soon realize that the glowing object is a pair of Apache-style moccasins, the extraordinary color coming from the pollen used to dye them.

LaVan has an entry in the tipi competition. If he wins, he will be able to buy enough gas to drive his family back to Arizona, where they live. I show him a petroglyph I photographed in Ontario that is unfamiliar to me. As he explains what it means, he plays a tape of a Hopi funeral ceremony, in which participants spend days recounting their origin from a swamp reed. As each generation passes, the story is recited by the survivors so they remember to pass it on to the next generation. But I'm not meeting with LaVan to discuss oral history. Instead, I'm here to talk about his progress in decoding petroglyphs of the American West and the relationship they have to Indian sign language. While LaVan was completing some fieldwork in Arizona, I'd been planning a trip to the Gaspé Peninsula in northern Québec to research the Mi'kmaq (pronounced ME-kmog) pictography, the "Rosetta Stone" for which is a Bible, translated by seventeenth-century missionaries into Mi'kmaq. I have a lead in Restigouche, a sleepy little town on the southern coast of the Gaspé where the native cultural center is located. This Bible is a word-for-word translation from French to the Mi'kmaq pictographic symbols and could be key to understanding the relationship that Mi'kmaq symbols have to symbols used elsewhere in North America.

We begin going over LaVan's fieldwork, and he pauses, midsentence, to hand me a stack of photographs. They are of one of the panels (which is what a series of petroglyphs on the same stone face are called) he is studying. We lay them out on a lawn chair and bend over them. LaVan traces every line with his ancient furrowed fingers, the texture of which mimics the sandstone in the pictures. His hands are like continents, moving tectonically over fading stories left by ancient strangers. It is almost as if he is aging with the rocks, becoming timeless, becoming the sandstone itself.

I soon realize what he is describing. This ancient panel is essentially a map of the southern portion of North America. I hardly breathe as he points out the geographical features and symbolic descriptions. Clearly, it was not Erich Von Daniken's aliens who recorded these images, but highly aware and cosmopolitan people with a deep sense of place, extensive and sophisticated communication capabilities, and an intricate understanding of the geography of America.

We then begin discussing the final project, which is to be an electronic dictionary of North American pictographs—their variants, contexts, and meanings. He drags out his old Mac Plus in which he just installed a bigger hard drive and shows me his progress in developing the database. He has not gotten very far. Most of his information is still in enormous three-ring binders, which consist of tens of thousands of pages of handwritten field notes, sketches, and photographs from the panels he's investigated.

The urgency of this project cannot be underestimated. As LaVan sees it, most people fail to appreciate that before speech-based writing systems, there were writing systems based on sign language with the capacity for nearly universal communication. The medium of choice was stone. This was the language of the Ice Age: a time of rapid climate change and the need for a durable and flexible way to communicate among many nomadic groups. With the migration of modern humans throughout the world, the rocks bear this language on every continent except Antarctica. We are haunted by images left by these Ice Age people wherever we go.

Preserving and documenting these sites is increasingly eclipsed by a hunger in us to dominate and manipulate the landscape. The very impulse and adaptability that sent us on these journeys, molding the land wherever we went, is now systematically erasing these pictographic documents. The destruction of these records is eradicating the history of indigenous peoples. Such pictographic documents store the daily culture, the epic migrations, the sacred traditions, the stories, the poems, the very humanity of all that preceded the pantheon of modern world cultures that now converge on North America to claim it as home. This cyberdictionary would definitively proclaim to the world that a universe of people as advanced intellectually, culturally, and socially as any other in the Western world thrived for thousands of years before European exploration. It would also lay the foundation for recovering and coherently piecing together this history so that it could be preserved before its extinction.

Although pictographies rely on sign language rather than speech, they are no less capable of expressing complex, abstract ideas. But they do require a different approach to being read than the written word, and they must be read, not just viewed. Writing is a linear "trail" of words. To understand a sentence, one must start at the beginning and end at the appropriate punctuation, as with the "flowing creek of speech" it represents. The written word is essentially a form of speech recording; temporal in nature, it must be moving forward to be understood. All modern languages are written this way. As with speech, the written word is very effective for telling a story as a sequence of events, but is cumbersome at describing spatial information, such as appearance and location. This information must be inferred from the reader's experience when reading a description provided by the writer. If the reader's frame of reference varies too greatly from that of the writer, then the imagery invoked may be confusing. Likewise, the emotion and context of the original environment cannot be captured, so it must be explained or summoned somehow by the author through the writing itself. This is the craft of storytelling. When we write, we always know what we mean to say, but the reader does not, so we must explain it in a language that we hope most others can understand. This is a problem faced constantly by writers. Even Plato complained to Phaedrus that the written word was useless unless the writer was present to defend and explain her or his work. The other problem is that writing is bound to speech. It is therefore unintelligible by those who do not speak the same language.

Conversely, petroglyphs are spatially distributed. There is meaning in the location of symbols, the way they are drawn, and their relationship to each other: they cannot be read in a linear fashion like a book. Thus, a panel is "read" more the way we view a picture. Our eyes must view the whole scene and wander from object to object to form an understanding of what is being described. The important difference between a petroglyph and a picture is that there is meaning in the way the various symbols are juxtaposed; their size, the thickness and shape of lines, and even the surface of the stone are used as part of the story in a panel or series of petroglyphs. There is nothing arbitrary included in these panels, since the intent is strictly to communicate, not to embellish. They can even depict the feelings of the writer toward the subject. Since all of this information is viewable at once, the gestalt of the story being told is available immediately and unfolds simultaneously, similar to viewing a picture, albeit a heavily annotated one.

The way LaVan Martineau works is by taking the whole panel into his visual field, then examining the detail in each stroke—the proportions, size, and relationship between figures at constantly changing scales. Then begins the work of comparing symbols in a particular panel with those from other locations. He has spent much of his adult life transferring the cryptanalytic skills he learned in the Korean War into this spatial reading technique.

No Word for Time: Obsolescence

Before this linear versus spatial communication was studied in great detail, Marshall McLuhan (well known for his research on the relationship between media and society and originator of the phrase, "The medium is the message") used a theory of acoustic and visual space to describe the effect of technologies on culture. Visual space is defined as linear, procedural, and single threaded, like reading a book. Acoustical space, in contrast, is considered spatial, multithreaded, and simultaneous, similar to the way we might hear a symphony. Although the labels for this dichotomy are somewhat misleading, the concepts behind it turn out to be important to understanding how information affects human perception. "Tribal" culture, as McLuhan called it, relates to its environment spatially, relying more on intelligence received directly from the environment to govern behavior and less on the linear narrative that time and duration impose on nature through writing. "Literate" culture relies on duration and relates to its environment in just such a linear fashion. It imposes a narrative on nature. Time for "literate man," as McLuhan calls us in the West, is of paramount importance. Events are timed and analyzed. Life is logged and punctuated by the calendar and the clock. For Westerners, the hands of time are always present. But the Mi'kmaq have no word for time.

In my own experience, this conflict arose in a longstanding interest I've had in the relationship the Labrador Coast Inuit have with the Torngat Mountains (pronounced "toong-aught" in Inuktitut), a range of ancient and ragged uplands covering the northern half of the province of Labrador. I wanted to conduct some interviews with the Inuit to better understand their traditions and perspectives on these rugged peaks. At the time, the eastern slopes of the Torngats (where the Labrador Coast Inuit hunted) were being considered for a national park or nickel mining claims. I had tried, in vain, over a period of months to contact key individuals in the community and determine the best path of research. Finally, I dropped in on a Canadian journalist I knew for advice. Although he did not have much experience with the Labrador Coast Inuit, he had extensive experience with those of eastern Ungava Bay, on the western slopes of the Torngats. As we talked, he looked down at my bare wrist and asked me for the time. I replied that I wasn't sure, since I never wore a watch, to which he observed, "Well, you're off to a good start." He suggested that I go to Nain (one of the more northern communities in Labrador, near the southern edge of the Torngat Mountains) and live in a tent on the edge of town for a few months with no particular plan, except to spend some time in the bush. Eventually, curious children would come by and start asking questions; then adults would follow. Apparently, it would also help if I got into trouble in the bush, such as falling and getting injured, requiring rescue by an Inuit hunting party. In our entire conversation, all activities required to come into contact with these people involved relating to them spatially: not through writing, not through scheduling meetings, but simply through existing in their "acoustic space." As far as my day-to-day experience with linear time and visual space, although I don't wear a watch, if I lose my date book, it's a pretty serious crisis. Let's just leave it at that.

LaVan and I had to cut our conversation short so he could go make final adjustments to his tipi before the judging was to begin. One crucial aspect of tipi construction is the configuration of the gap at the bottom, which allows smoke to rise efficiently through the vent in the top. Aside from the obvious practical function, this has an important spiritual function as well. The rising of smoke is a way of communicating with the sky world, just as the church steeple is an antenna toward heaven. In spirit-to-real-world connections, the effectiveness of this communication might ultimately affect how well the people are heard. Authenticity is more important to LaVan than fancy design, so he wants to get it exact. He suggests we meet again after the dance competition, which is about to begin. On my way over to the dance platform, I stop at the vendor tent to look for out-of-print books. By the time I arrive at the dance venue, people are crowding in quickly, making it impossible to move until it's over.

During the dance, I realize that I've forgotten where LaVan and I are supposed to meet later that afternoon. It would be impossible for me to get to him or for him to hear me over the noise. I spot him on the other side of the platform and catch his attention, giving him the sign for location. He returns the sign with detailed directions to our meeting place without any electronic or verbal form of communication.

So how did such a practical, rich form of communication like the Indian sign language become limited to a few insiders? How did the resulting pictography grow extinct so quickly? As with all other forms of cultural archaeology, no one knows for certain, but everyone is certain they know. Urbanization and agriculture are the likely suspects. There were compelling reasons for humans to urbanize. Humans are ingenious yet frail, descended from small arboreal ancestors: not the hunters, but the hunted. According to evolutionary theorist Steven M. Stanley, as the climate changed and forests disappeared during the Ice Age, humans were cast into a grassland filled with lions and hyenas.[1] One can only vaguely imagine the predicament of these earliest "proto-people." Flocking behavior is a natural response of prey animals stuck on the ground, but with a constantly changing climate, nomadicism would have been necessary for finding shifting food sources. There was likely a strong desire to conserve energy, to find a place to settle permanently and call home, but as food ran out, seasons changed, and glaciers moved, so did the people. Smaller groups would have been more sustainable, since they could have taken advantage of sparser resources and been more flexible about when and how to move. A very large group would be less vulnerable to predation, but would have had a harder time finding enough food for all, and would have taken much longer to pack up their belongings when it was time to move on.

With a warmer, more settled climate and more plentiful resources, flocking behavior would have compelled groups to settle, increase in numbers, and quite naturally rely more intensely on speech. The problems of isolation no longer inhibited communication. The problem of having enough to feed a growing population was eventually solved by agriculture, a codependent response to mutually assured survival since we guarantee the success of those plants and animals we grow in order to have successful populations ourselves.

With urbanization, the problem shifted from "how do small groups who move around a lot and speak different tongues communicate and remember what's where the next time they pass a particular place" to "how do we keep track of whose sheep belong to whom, who said what about whom, who is in charge, how do we make rules that everyone will follow . . ."—in other words, how do we get along together and share resources, skills, and so forth? The pictographs were obsolesced as representing ideas and reused to represent speech. Society shifted from living in acoustic to visual space. This transition affected more than just writing; it affected the way people saw their place in the world. The importance of time, distance, order, control, sequence, dominates over simultaneity, holism, gestalt, synthesis, and creativity.

Arguably, the urban, literate mind-set has also prevented us from living sustainably. A dream come true can turn out to be the worst nightmare. The human brain, molded by ice with a fierce determination to find and exploit resources, figure out creative ways to use and manipulate those resources, and procreate as often as possible, never anticipated how the world would change once it could actually achieve these goals. The forager mind humans were equipped with, as intelligent as it was, does not act as if it ever intended to deal with problems of urban societies, specialization, population growth, and the technological complexity that evolved. Some researchers, such as Julian Jaynes, speculate that consciousness (as we know it) may have arisen from the confusion and chaos brought about by urbanization and intense human interaction.[2] One thing is for certain: speech-based writing, environmental and health problems, urbanization, formal agriculture, and even possibly consciousness all arose together over a period of a few thousand years as the glaciers receded and the climate stabilized. Since that time, nomadicism and pictographic writing have been systematically abandoned. The last vestiges may remain in a few isolated aboriginal Australian societies, which are under constant pressure to Westernize.

Blood and Television: Retrieval

The intense focus on the linear and analytical has led to an explosion in technological innovation and miniaturization in order to virtually eliminate traditional dangers such as lions and hyenas. Humans have created many of the problems associated with visual space: alienation from nature and each other, nervousness in dealing with the increasing rush of information technology, ethnocentricity and world wars, environmental devastation, monoculturalism, and unsustainability—to name a few. Technology highlights these problems, creating a dualism between the way we behave and the way we intellectualize our place in nature. We yearn for integration, we want to be needed, to feel part of a community, to feel at home, to be part of something bigger than ourselves, yet we pick nature apart and impose precision and intense specialization. We look at our history, but do we really see it? The increasing speed we are moving at is tearing us apart, yet we look back only to admire the patterns.

There is increasing evidence that perhaps the direction and intensity of technological development (in particular, information technologies) are devolving away from the linear thinking that created it. Just as magically as we supplanted the sign language and pictographs with writing in our urban and agricultural centers, we may now be abandoning the printed word for images in the "global village." The twentieth century was the bloodiest in history, but red was not the only color we saw. Visualization and visual communication came to dominate early on with the development and proliferation of images in television, film, and publishing. We not only experienced the bloodshed; we recorded it in vivid, colorful detail. Arguably, civil rights and the environmental and antiwar movements of the 1960s were propelled forward through television and print. Perhaps this is because the entire planet is beginning to look like the same anthill. We run into more conflicts over the same resources and differences of opinion as to how to use them. Thus, it may have more to do with how technology has brought us into increasing contact with one another and these resources.

The twentieth century came to a close with the use of sophisticated visualization in nearly every form of inquiry, analysis, and design. Networks of remote sensing satellites cast a nearly unblinking eye on the planet, just as we paradoxically obscured our view of the heavens with the light and smoke from our cities. In that century, we took 50 billion photographs a year, and the Web gave us a way to share them with everyone on Earth. Image manipulation through programs like Photoshop allowed us to manipulate image content to tell stories the way we wanted to rather than allowing them to remain a snapshot. The medium was not only the message, but became the way of life: the lens through which we perceived ourselves and the world around us.

Why Johnny Doesn't Read: Reversal

Teachers lament that kids don't read much anymore, but the reality is that neither do adults. It is not uncommon to come across a household without a single book or magazine yet a television in every room. The rest of the time seems to be spent surfing the Web. Images are increasingly used to portray ideas from the simplest computer icon to the most complex scientific data set. In the digital world, text supports the images rather than images supporting the text. As digital television, digital photography, and broadband data communications get increased market penetration, perhaps the need for text in everyday communication will eventually cease. Images will be manipulated with such ease that today's photographs will seem as primitive as petroglyphs. (How crude and barbaric, after all, is capturing gelatin and dye images for which we must boil the skin and bones of animals and impregnate the resulting goo with toxic chemicals.) Teachers will stop asking why kids don't read anymore and start asking what the meaning is for the way they choose to juxtapose imagery in a specific way. Perhaps then the literate West will be ready to read the rocks, will understand why this intricate form of communication was used, and its value in understanding the history they hold. But by then, most of the petroglyphs will be gone and, with them, their stories.

As the Earth has moved into a new millennium, literate man clings to time, scarcely pausing to look at the now or the historic. "Digital man" may care even less about casting aside his legacy. But while we may not care about history, nature does not let us forget the past. Even as digital beings, or perhaps because of it, we will still have the brain of the nomad. The interplay of memory and the senses that allowed us to survive the Ice Age and populate the planet will compel us with a desire to see, hear, taste, smell, and touch even our own synthetic worlds. Only now *we* are the geologic force that rapidly changes the climate and shapes the landscape, demanding a flexible, universal way to communicate. The durability of this new language lies only in its verisimilitude and integration into our individual realities as its real form is carried away by the pipe smoke of anonymity.

We have molded the stones into new rocks that speak: the personal computer, the telephone, the television, the fax machine. Perhaps McLuhan's observation was correct that our "electric networks" are an effort to return us to those tribal roots, returning to the rocks, in which we will peck and paint our histories. When LaVan and I were trying to find an effective way of portraying petroglyphs on computers, we failed to observe that perhaps the purpose of these images would translate on its own. Perhaps this ancient visual language is being recapitulated in the World Wide Web and other electronic media.

The ephemeral networked nature of the Web represents our culture spatially, through images at the expense of the written word, allowing its rapid extinction in pursuit of change. Web pages come and go quickly, many of them scarcely exist, and then, "Not found. The requested URL was not found on this server" appears. Electronic networks rely heavily on the social networks that support them and provide their content. Likewise, no one would have bothered to peck away at sandstone or press text into a page if it was never going to be read. But the nature of computer-based imagery is that it must be reconstructed in its environment for each viewing. It is purely allographic. The technology "performs" the piece precisely the same way each time it is asked. Computer technology is obsolesced in a matter of years, sandstone in millennia.

Where will this lead? The human impulse to communicate and create has propelled us into a world of increasingly complex electronic and social networks, but as other cultures come to dominate us, will this sophistication be forgotten? Will we miss the opportunity to use this new acoustic space to transcend linear thinking? Will we be understood as a bunch of primitives who crowded into cities, laid waste to paradise, and wiled their lives away worshiping glowing stones, advertising slogans, and consumerism? Or will those who analyze our culture, methodically document our images, and slowly decrypt them understand us as sacrificing everything to build networks in a search for truth and meaning, to build a global civil society? Perhaps our images will simply evolve into icons that will be strung together into a new literate culture, which will evolve in visual space to abandon nature once again—to abandon the tribal, spatial roots, and begin a new age of "enlightenment."

But for now, the rocks continue to be weathered, blasted, bulldozed, and buried; the written page is browned, cracked, burned, and landfilled; Web pages are deleted, corrupted, moved, and unlinked—as the sky turns from pale to blue to pale of dusk, as the Earth turns, as powwows come and go, leaving empty fields punctuated only by a few overflowing trashcans. Dust swirls around in the late summer wind until rain turns dust to mud, flowing into the river, flowing away from the powwow grounds. Now it is a silent place, where a man who talked to stones once held a pair of Apache moccasins glowing in the sunlight, glowing in the eyes of a traveler.

Afterword: Eulogy

After the powwow, LaVan and I continued working together, building a library of existing knowledge. As he continued his ambitious decryption and fieldwork, I sought out and recorded these mysterious symbols where I could and requested funding from government and private foundations to further LaVan's work. But LaVan Martineau was a free agent, not affiliated with any university. Most foundations found it difficult to understand his work. My pleas for support went largely unheeded.

To me, the most frightening prospect was that of losing this important knowledge without a "backup" copy anywhere, whether through the continued destruction of petroglyph sites around the countryside or a mishap in LaVan's library resulting in the destruction of his precious loose-leaf binders. And worst of all would be the loss of the mind that held this knowledge. Every furrow, age spot, gray hair, clearing of his throat reminded me of black arrows creeping inexorably across the clock face. I imagined the death of LaVan Martineau as the burning of the library at Alexandria.

Within a few years, which seemed to me like seconds, these arrows caught up with him. At the turn of a new millennium, the man who spoke to stones and carried their stories lay in a casket in the Shivwits Tribal School as friends paid tribute with the Bird Dance. "Fly away, fly away, fly away home," they sang as they danced birdlike around LaVan Martineau's coffin, surrounded by his belongings to accompany him on his journey.

The rocks have fallen silent once again. Others will have to take up LaVan's torch and carry it into the cave, to read the rocks once more, to see where ancient strangers pecked figures or touched ochre to stone. At LaVan Martineau's funeral, no one asked why kids don't read anymore. Transformed from the wise elder to a winged spirit, free of suffering, LaVan took some of our history with him as he began to navigate among the constellations. On that day, the construction crews that were building the highways, the subdivisions, the mines and dams must have paused, even if just for a breath. It is not possible that such a passing does not cause our grip to loosen on the tools that worry away the past. But with just as much certainty they must have resumed, compelled forward: nomads traveling ceaselessly through the pristine wilderness of time.

Notes

1. Steven M. Stanley, *Children of the Ice Age: How Global Catastrophe Allowed Humans to Evolve* (Harmony Books: New York, 1996).
2. Julian Jaynes, *The Origin of Consciousness in the Breakdown of the Bicameral Mind* (Houghton Mifflin Company: New York, 1976).

Peter Goin, *Mannequin Prostitute* (detail)

From *The Cane Flute*

Kristjana Gunnars

Sometime in your life, you wake up and know it has become necessary to change things. It has dawned on you imperceptibly that you have slept too long. The sky is already pale blue. The mountainsides are pink and low blankets of cloud are rising from the great forests. You know not just one or two things have to change. It is your whole life that must be altered. You have gone down this road too far. Your mistakes have compounded themselves and you are trapped.

It is a way of standing on the edge of a continent. In front of you there is only ocean. The ocean goes on without stopping. If you continued straight on that sea, you would be on an island in the South Pacific. Somewhere in the presence of palm trees and bamboo groves. You would have pineapple and coconut for lunch. You look at the sand and the water and think it would be nice to go there. That would be a nice change.

It seems the past is simply time petrified. What has happened before is stopped in its tracks, there forever frozen. You think about your own past. There are some things you are sorry for, but you know you cannot change them. Some things you are angry over, but you also know there is no one who can give you closure. You are left to your own devices. All you have is your own intelligence.

On such days the sun taunts behind a thin cloud cover, but does not come out. The light edge of late morning remains gray. The fall leaves are yellow, brown, ochre, russet, the sky in the far northwest is tinged with lilac as late into the day as noon.

The radio plays bombastic symphonies all day, with trumpets and cymbals and percussion rolling. There are machines tearing down cedar forests with their loud growling noise. You know there is an end in all this. The life of the forest ends and begins again. Something else takes its place. You see something inevitable in the altering landscape. You say to yourself, you can let it go. I have been fighting all my life. I will fight no more.

Sometimes the layer of white cloud distends into a sheet of bone-colored smoke. Then the smoke lifts and reveals the eggplant-tinged hills that layer themselves into the water below. A small gray boat cuts a half-moon across the bay. It is the smoke of autumn, when mornings are cool again and the house has a damp air about it.

I find myself settling in for the winter rains. I am bringing up firewood and piling it beside the wood-burning stove. I save last week's newspapers for kindling. All the yellow, orange, and red flower petals are drying on the wicker plate. I have plans to keep the petals for potpourri all winter long. A plate of seeds is on the deck, for stray blue jays still around. On the grass outside my window, two deer are eating up weeds in the kinnikinnick. When I look out, they stop and stare back. But eventually they forget about me and go into the flowers, looking for roses. When I turn, the bulky, flabby body of a big black bear slouches into the bush.

For all the learning we supposedly accumulate over a lifetime, it remains a reality that we have very few moments when we think we know something for sure. Most of the time we only think we are sure. But real conviction, the kind we would stake our lives on, does not come easily. Such visions have to be lived. Not earned, not learned, but rather come to through experience. Experience that freezes what it knows.

There is one thing I know for sure. I know there is such a thing as grace. Grace does not come when you want it but moves of its own accord alone. It is a gift. You have grace without earning it and you lose it without deserving to. Grace has the air of morning fog. It is there, you see it everywhere, then it lifts away.

I know there are times in life when grace departs. You are left without, bereft. There is no inner voice. Life has no symbolic dimension. Everything is flat and you don't hear anything resonate. Then one day you recognize something has come back. There was no flash of light and no special sound. Over time you come to realize that grace is with you again. You have direction. Suddenly you see how things, small and insignificant, have meaning.

I get my material from around me, Abbas Kiarostami said in an interview. *I believe that any artist finds his material from what's around him.* In Kiarostami's films, the protagonist is always running from one place to another, but it is never forward. He is endlessly traveling, zigzagging back and forth. He will never reach his destination. While he is on his way, he encounters strangers who tell him the story of his life, his country and his times. But they tell this to him in bits and pieces. No narrative is ever complete. No story hangs together.

It takes as long to write all this as it does to make a film. The location has to be found. The light, the time of day or night, the disposition of the actors. A whole crew is needed. The speakers and readers have to come forward. It is just like searching for the camera operator, the set and lighting designer, the editor, the director. And after, the publisher, the distributor, the reader. All this is not done in a day. The sun rises and sets many times while the words drag themselves up from the beach, where the small pink and purple crabs claw out of the water onto the shale. It takes time to bring this fluid thing into the steady gaze of a past, and make it the same thing you see every time you look.

It seems to me now that Iranian films have a lot of tragedy in them, sadness and depression. But the love people have for one another comes through like a fifth character and holds it all together.

Winter rains become something else as the cold air takes over. We come to live in a kind of isolation. Sometimes we are snowed in up here on the mountain. Down below life goes on as before: the shops open their doors, their window signs turn around to spell "Open," the construction crews return to the highway, women dressed in down-filled parkas hold up signs that say "Slow." But up here among the leafless alders and the Oregon juncos, we are stranded. The roads are slippery with ice. Along the roadside there are small hills of snow where the plough has pushed its debris.

We awaken to the sound of a snow scraper. We look out the window, and thick, fluffy snow is floating soundlessly from above. The branches on the deciduous trees, so gray and unmoving the day before, are lit up with layers of snow. The forest outside has become thick, has taken on the coating of a mother, hugging herself in the soft dawn. On such days we decide, without regret, to stay home. To light the fire. Read a good book, play some music. But this time when it happened, I knew I could get out. I could go into the village. Isolation is not inevitable; I just used the snow as an excuse. I said to myself, we are snowed in. We cannot get out.

In A. B. Yehoshua's story "The Continuing Silence of a Poet," the poet who has stopped writing altogether, never to start again, tells us: *And my silence was accepted—in silence.*

Winter, barely noticeable, turns itself into spring. Quietly. Effortlessly. There were sudden downfalls of snow, peculiarly frosty mornings that melted after breakfast. But now it is rain again. The railings are wet. Drops line up under the patio table. There is a thick fog that shrouds the cedars in white cloud. The more distant trees are barely visible and stand there as if they have yet to depart with a dream.

On such a morning, I am reminded that now I am exactly where I should be. How seldom one can say that, I have often thought. How many times have I arisen to the thought that I should have been somewhere else? But the cedar trees are my sentinels. The juncos are my company. The sterling jays are the perpetual chorus of my existence. There is peace. Here in this part of the world, on my single hook of land that juts into the Pacific waters, there is quiet peace. Of course, this feeling, this too will turn into something else. Should I mourn the loss of peace finally found before it is gone again?

This year I have decided to celebrate Ash Wednesday for the first time, because it is there. The day has been commemorated for centuries, and it fulfills a purpose which I have overlooked until now. It is the purpose of repentance. It occurs to me we have so many explanations for why we have done things, why we have acted as we did, and why we will not acknowledge our actions as wrong, that we have made no room for regret. Real repentance is not of our time. Repentance is not even a religious matter. It is a matter for the conscience. When we have done something we know in our hearts was wrong, and then try to justify ourselves although it's not necessary. We don't need to secure justification but we do need to know how to repent. To be able to say, "I did this and it was wrong of me."

All of this came to me as I was reading Doris Lessing's novel *Love, Again.* It was not the novel itself, the writing or the story, but just a moment's thought the narrator has, and then brushes past. The narrator says of our wrong actions, in the roundabout way of how we are driven, that *there seems to be a rule that what you condemn will turn up sooner or later, to be lived through.* And then I remembered something Kazuo Ishiguro said. He said we can make lists with pros and cons for our decisions, and we can think them through as carefully as we like. But in fact, the important decisions in life—whom we marry, where we live, what jobs we take—are not made so logically. The big things we do are done because of our needs. We are driven by primeval forces to do certain things, and they are not always good.

I do not think, when you set out to change your life, that you can change just a few things. Maybe the place you live, or your job, or even the one you have married: and then continue with the rest. I think you will have to change everything, from top to bottom. To "fumigate" your entire life. To put all you have lived to this point behind you: and start all over.

It is better that way. Then you will not take the debris, the dust, of your old life into your new. You will be able to change freely, without encumbrances. If you hang on to everything that was before, you hang on to the inability to change. And when you look for change, you want to move to something better. Not something worse, or even analogous, but better. To change for the good.

Looking at the flame in the pine-green morning candle, I know that change has a spiral form. If all things are moving around, the way the seasons revolve, they are always returning in different forms. Life can be like that.

While this is happening, as winter turns into spring, I want to let the stillness of the morning enter this space. The mountains and the ocean are being gleaned on by the morning sun, which is trapped behind thick layers of cloud. The vista is the color of plums in a bowl, piled high one on the other: some are almost red with ripeness, some are green from being too young. A mountain of plums which the light of day has begun to shine on. I see it through the window.

Sometimes I think it is possible to get over the past, to get to that better place, only through ritual. That it is impossible to speak to the soul in any other way. The soul is not intellectual or logical. It understands the symbols of our lives, that's why symbols are there. To tell the soul to forget, to absolve, is nearly impossible unless a ceremony is performed. What the ancients used to call "placating the gods."

After thinking this, I tried to find a way of bringing my mother, so long passed now, into my daily life. Because she is not forgotten, she lives here. She floats around my house like a fog, and is never acknowledged. No one speaks to her. I got a bright red candle in a glass cup, and put it in a prominent place where the kitchen meets the living room and the dining area. Sometimes I can light this small candle and watch it burn. It is her candle. I see the little flame reflected in the black window in the night. Her small flame. In a strange way, she is changelessly alive. I wonder if that is the difference between those who pass away and those who are here. We still change. They do not.

Nothing is as it seems. Or so it seems. Not even this sentence. Jack Kerouac wrote in *On the Road* that he, the narrator Sal Paradise, is watching his friends Dean and Carlo face off, trying to be honest. They try to be more and more honest, saying a more honest sentence than the one before, but there is always something more honest to say after. So they go on, all night, stripping layer after layer. Finally Sal says: *That last thing is what you can't get, Carlo. Nobody can get to that last thing. We keep on living in hopes of catching it once for all.*

Then Sal is in California and everyone is rushing around, trying to get to the thing in front. *Everybody was rushing off toward the farthest palm,* he says, and the streets are lined with them. It occurs to me they are not running forward. Kerouac has described another zigzag, another kind of circle.

It darkens early now. The autumn colors begin to show themselves. Outside the window, I catch the first glimpse of a bright red maple leaf. The maple leaves are very large this year. They seem huge. The Steller's jay is back, screeching the way it does in winter. My window is open, and the bright steel-blue bird can hear music from my stereo. "Silk Road Music," in a high pitch, just the tones they like. The jay stops moving and stands on the porch table, stock still. Listening.

The moon is out now in the western sky. At dusk, the full moon shines starkly, illuminating everything in its neon glare. There is always this interstellar feeling in the moonlight. Such a pale and lonely glow. Melancholic.

In the moonlight of early dawn, I am thinking of ashes. How it is possible to return again from ashes. After the city of your life has been sacked, every building you constructed is in ruins, you can clear up the debris and start again. Your city was beautiful, you feel a great loss. You have forgotten that it is possible to come back. And it is possible to remake the city, so it is more beautiful than before. First you look at the ashes and you feel it will always be like this. Always in ruins.

I think it is all right to be disoriented for a while. Disorientation is part of the picture. So you wander about the place, listening to the time pass. If you try hard, you can hear time go by. You can see it when the light changes. How the pale, white light takes in some aquamarine, then a slight rose, a blush. That will be time passing. You can hear time moving on—in the rain on the skylight, in the branches hurled against the house by the wind.

I remind myself that if you give everything away, you no longer have it yourself. And even if the things you had before are still there—the teapot, the bookstand, the shawl on the sofa—you may have given away the spirit of the place. It takes a while to have possession of your own life.

There must be quite a few things a hot bath won't cure, Sylvia Plath wrote in her novel *The Bell Jar,* when her protagonist was sinking into depression—*but I don't know many of them.*

Browsing in bookstores is a minor hobby of mine. People who buy things on-line miss the pleasure of wandering about among myriads of books, sitting down in big chairs and leafing through them. Then taking some special find home—a book you did not know existed. Holding it in your arms on your way out the door, looking forward to curling up with it later. What adventure may be awaiting in this book, you never know. Perhaps it will change your life. Perhaps not.

Bookstores are good places to dream. The books are lined up in the stacks. It is usually quiet; people walk about respectfully. There are hushed voices. Those who are deeply engaged in their browsing throw glances at each other with secret understanding. Book browsers are spies in the land of dreams. In bookstores I think thoughts like: accepted wisdom is not always wise. Such as: don't judge a book by its cover. I usually judge a book by its cover. If the cover is beautiful or intriguing, I know the publisher cared about it. If the publisher cared, so should I.

Accepted wisdom, such as: don't let them change you.

When the twin towers in New York City crumbled against terrorism, that was the first thing people said. They said: go on as before, don't let this change you, because then they win! But I think we should be changed by events like that. We should absorb what has happened. It happened for a reason. We should be shocked and hurt and made thoughtful. We would be better if we changed and came out more beautiful than before.

In late October, the winds arise from the sea. Winter gales set in. Trees begin to hammer the sides of the house and the windows in their frames. The dawning day is blustery, the clouds billow. The water looks like a white, neon sheet under the heady canopy. All is alive. I step outside, and the smell of wet cedar stings. Big, yellow maple leaves fly heavily across the path. At such times I feel like walking against the battling winds.

In the early morning, I listen to the floating sound of the koto and the shakahachi. The players improvise a meditation in sound, and while I listen, the leaves rustle on the branches. A dawn wind is agitating the hour. Once again I have opened a book of the poems of Rumi, almost by instinct. It does not matter why or what I think I am looking for. What is extraordinary in this literature, I find, is the strange combination of passion and tranquility. How can these exist so closely together? The rejection of the world and the love of the world, together like the Wolf Brothers circling the moon forever.

In the last stanza of a poem called "On the Threshold," the speaker says he wants to sit quite still, in between two worlds, and listen to the silence. I am thinking, perhaps if you take away all the noise of the world, you can still not find silence. Perhaps complete silence does not exist where there is life. It may be that all things sing, and when everything is quiet, the song inside you comes out and fills the void.

In the summer I watch the dawn come in. Daylight enters so quietly, a soundless scape of lilacs and pinks and purples. When there is no wind, the world seems the most peaceful place. I wonder, Why did they go looking for Paradise when it was right here all the time? Do we chase around the rainbow for what we already have at home? Then I remember something Thomas Merton said. He said, we have to create this thing we think of as Paradise. It was not there before. It is a memory we have, lost in the archives of time.

The sun filters itself through alder leaves, and the world is green, orange, avocado, bright neon lime, in speckles and patches. I end up thinking about words and how they lock people in. What the words of others do to confine you inside their lock-box. The prisons of that-which-others-say. How words, once spoken, can take away the movement of time and the fluidity of life.

I do not know why I am thinking this. Be careful what you say, because it may kill someone.

When Malika Oufkir was asked what she had learned from her experience of being held with her family in prison for fifteen years and dangerously escaping out of Morocco, she said: *If you can save someone's life, then you must do it.* She did not mean a daring act, like on a trapeze or high wire. *Maybe it is only a word,* she explained. Maybe only a word you say.

I continue to come and go all this time. When I arrive at my home in the country in mid-June with my flight bag, knowing I will stay for a while, I notice the coolness. It seemed unusual. This is no summer—not the kind of summer I remember. I think everything is unusual now. This will take time getting used to. I tell myself it is all right to take the time. When you have been far away from yourself, it is all right to take time to return.

It could happen any time. Anywhere. I know this is true. It could occur, that one day you realize you do not remember who you are. You look around at things that should be familiar. The blue and gold teapot. The brass pitcher. The cedar tree protruding into the view. But they seem unfamiliar.

I put the bag in the hall and open the window shades. I will live here again, watch the seasons go round, observe the birds and animals and trees. But it is not the same place. I am turning into someone else. Another form of myself. I will watch that too. It is something unknown, and it will become clearer as the fog lifts.

New England Ghazal

John Canaday

In the beginning were the Words of God, disguised as stones:
like hard, black pupils dropped into the faithful's eyes, these stones.

Waves hunched in worship shake the granite shore beneath my feet
as once it shuddered under the soles that colonized these stones.

Salt of the earth, they said, "Let nothing grow upon this spot
till Hell silts over. Let them lie among blowflies and stones."

No schist for me, no strata resurrected from dead tongues.
I'll cleave to coal and shale and strive to anglicize the stones.

Cursing a blue streak when his plow beached on a granite spur,
the farmer wiped his brow, letting his sweat baptize these stones.

"Fuck you, you fucking fucker," froths a four-foot grade-school kid,
thrilled to outdo his friends, who've grown curt, course, streetwise, like stones.

Praise limestone. Mouth a marble chip. Haunt dikes, tors, sills, and crags.
In drought, clay sheathes its softer self and identifies with stones.

Such meditation on the inner life: the CAT scan's Ommm;
unearthly images of what metastasized as stones.

Who doesn't long to blame someone for our infirmities?
I turn to God and nature, but their alibis are stones.

The Jackal sings his privy business to the world at large.
Hungry, unfit to kill, he grumbles lullabies to stones.

The Wings of the Wind

Floyd Skloot

> *He came swiftly upon the wings of the wind. He made darkness his covering around him.*
> —Psalm 18

My son Isaac, adopted at twelve weeks of age, turned twenty on September 15. That day, he measured exactly seven feet five and three-quarter inches tall. When he was home last summer, I stood beside him posing for a photograph and my head came up to about his pancreas.

Isaac outgrew Loretta when he was eight and me when he was ten. But until he was sixteen or so, he loved to walk between us holding hands, towering over us like a spike on a graph. He would giggle at our reflections in shop windows and at the looks on the faces of passers-by. Isaac weighs exactly what my wife and I do combined, 290 pounds. And as you can see, I'm accustomed to using precise numbers when I speak of my son.

Loretta and I got him from an orphanage in The People's Republic of Romania at the zenith of Ceaușescu's power. All we knew was that he had been born near Sibiu, a small industrial city north of the Transylvania Alps, his mother had been a textile worker, and he had such raging impetigo around his nose and mouth when we first saw him that it looked as though someone had drawn on a bright red clown's face.

We named him Isaac after my father, but the name would have had a nice resonance for us anyway. According to Genesis, the biblical Isaac was born when his parents were quite old and thought themselves well beyond child-bearing age. Same as us. He was also the fulfillment of a promise, which felt right in our case as well, a kind of entitlement due to a couple who loved as deeply as we did. Then what does God do after giving Sarah and Abraham their son? Right. No wonder Loretta and I were overprotective, despite our Isaac's size.

He was virtually silent during the entire plane trip from Bucharest to New York. He slept for long periods, waking with a soft whimper wanting to be fed, then sitting in his carrier with his eyes wide open, hands moving vaguely in front of his face as though warding off something strange and unpleasant. His smiles seemed wholly inward, prompted less by anything he saw than by forces going on inside his body.

Probably you've heard of him. Isaac Berg, the great basketball star. Most people call him Ike and the press dubbed him The Ikeberg, a huge mass afloat in the middle of the lane. They say he may be the best collegiate center since either Wilt Chamberlain or Kareem Abdul-Jabbar or Bill Walton or Shaquille O'Neal—depending on your era—and maybe the best big man ever. In his freshman year at the University of Oregon, Isaac averaged twenty-six points, fourteen rebounds and six blocked shots a game, leading the Ducks to the NCAA Final Four where absolutely no one had expected them to be and where they had never been before. People in and around the game thought he would leave school then and make himself eligible for the pro draft, but he stayed on and had an even better sophomore year. In a press conference last spring, to stop all the speculation, he announced that he was staying in school for his final two years. He wanted to get his degree, help the team to its third straight final-four appearance, and play in the Olympics as a true amateur before basketball became a job for him.

I hope you got to see my Isaac during that press conference. Except for his size, everything about him said Gentle Scholar, said Books, said Soulful. For the first time, his nickname seemed to make sense, to reflect the truth that so much of Isaac was hidden below the surface regardless of how much there was above it. After every question, he looked down as though to collect his thoughts, smiling shyly, humble, unwilling to seem the smart-ass like so many other young athletes. It was the proudest moment for me, prouder than all his on-court honors, prouder than the night he scored 67 points and grabbed 28 rebounds against U.C.L.A.

The name Isaac Berg, I would be the first to admit, fits a little strangely on him now. We had him circumcised when he was an infant, brought him up within the Reform movement as a moderately observant Jew, and kept telling him that he, of all people, was truly one of the Chosen. Of course, now he looks about as Jewish as Vlad the Impaler, but who knew that when he was six years old? He was a dark infant, had huge hands and feet, a mop of curly black hair and deep-set eyes that followed me everywhere I moved, and while I didn't think he was necessarily of Jewish parentage, I didn't imagine he would look quite so gentile either. His features became enormous, everything growing at an accelerated pace, and, with his basso profundo voice thrown in, my son could be quite terrifying to encounter. Which is so ironic, since I believe it worries him to box opponents out for a rebound lest he accidentally crush someone.

At his Bar Mitzvah, Isaac stood six foot five and towered over the rabbi. Most thirteen year olds have to stand on a stool to see over the podium when they read their Bar Mitzvah portions to the congregation; Isaac, clean-shaven but heavily shadowed with stubble by 10:00 A.M., had to stoop so that Rabbi Herschorn didn't hurt his neck looking up to bless him. As Isaac carried the Torah in a slow march through the congregation, some of the little old men reaching up to touch the scroll with the hems of their prayer shawls suddenly backed away as though seeing a Cossack on horseback. But he was oblivious, blissful at being a newly consecrated member of the tribe, his face literally glowing as he passed into light that poured through the stained glass window at the rear of the sanctuary. That night, at home after his party, Isaac could not stop talking about feeling hugged by God—those were his exact words. He had been embraced in that light, taken over, and felt himself to be loved and protected forever. It was a feeling I have never had myself, even at his age, and I remember hoping that whatever caused him to lose it would not be too great for him to bear.

Like so many extremely tall young people, Isaac had his social difficulties, especially as an adolescent. He didn't join clubs, didn't have any close male friends, didn't date much. The first time, near the end of his sophomore year, he asked me to drive him and the young girl, a foreign exchange student from Paris named Laura Quost, to the downtown cinema where a French film was showing. I dropped them off, went by myself to another film in a theater across the river, then picked them up at the Metro on Broadway, where they had gone for snacks and sparkling water, four hours later. Isaac's relief when he saw me walk through the restaurant's doors was so palpable that I felt like crying for him.

He knew he was strange-looking and, I think, secretly agreed when kids called him Frankenstein, called him Moonman, Geek or Bronto. But that all changed half-way through his junior year in high school, when Isaac's talent as a basketball player asserted itself and he became a hero over the course of one frigid Portland winter.

I will never forget the silence that fell over the gym the first time Isaac's rage was expressed on the court. The silence only lasted for two seconds, a kind of collective stoppage of breath, before turning into something like joy as the fans erupted in whistles and wild cheers. But it was those two seconds I will never forget. Isaac had been taunted by the opposing fans and, worse, by players on the other team, throughout the first half of the season's opening game. He had refused even to consider joining the team during his first two years in school, refused to be seen with a basketball anywhere except on the driveway beside our house, and as the game progressed he had played mildly, passing the ball off whenever he got it, sticking up his arms but not moving aggressively to block any shots, reaching for rebounds but not boxing out or jumping. It seemed as though he was afraid to stumble and look awkward, though I knew it had more to do with his fear of causing harm. As the third quarter was drawing to a close, I saw his eyes narrow and his nostrils flare and I felt that something had snapped in Isaac. After all those years of being razzed, of trying to act small, denying his essential isolation, he had finally grasped some essential truth about his situation and reached an instantaneous decision. He moved into position at the top of the key and raised his arms, calling for the ball. The point guard, as shocked as I was to see Isaac asserting his will, bounced the pass to him. Isaac planted his feet, spun to his left and took one incredibly long step toward the basket. With the ball securely stuck in his right palm, threw down a slam dunk in one great windmilling motion. The ball tore through the net and bounced back up off the floor so high that time seemed to stand still as everyone watched it reach its apogee before reacting to what they had seen. Isaac had dunked from the foul line, moving through the air with such power, authority and grace that he looked like a seasoned professional. Or a prehistoric bird riding a zephyr. Back on the ground, he stood there glaring into the middle distance while the gym filled with noise. Then his eyes changed again, found me where I sat at midcourt as he trotted back on defense, and the expression on my son's face was a terrifying mix of triumph and grief.

Believe me, there was never a spoken plan. However, Loretta and I understood that Isaac intended to buy us a new home on Lake Oswego as soon as he signed his

first professional contract. He thought we might also like matching leather recliners for the new living room, a big screen television on which to watch his games, a dark green Lexus, a summer trip to Israel after the basketball season. Where he came up with Israel, I don't know; my fantasy has always been a month on the beach in Rio.

This idea of Isaac's was something we knew from hints and suggestions he would drop into conversations. We'd be sitting in the living room after dinner, all three of us immersed in our books, and Isaac would suddenly wonder if it would be nice to listen to some Mozart right about now, maybe that Piano Concerto in C Major that we all went to hear last year in downtown Portland. Be nice to savor the crystal clear sound one of those new CD players is capable of. Then furniture and electronics catalogues started to arrive for us in the mail after his freshman season at Oregon. We were suddenly on the mailing lists for travel agencies and fancy automobile dealerships. Taking care of us in this way was probably the only thing that made Isaac hesitate to turn down the professional inducements and stay in school till he graduated. I'm glad he did.

These dreams on our behalf were pure American dreams; they certainly weren't Romanian ones. So equating personal success with waterfront homes or luxury cars is not in our genes after all! Tay-Sachs disease is in our genes, a hundred times more than in anybody else's, but not the need for a big screen television. Excuse me. I've been a bit emotional lately.

Although Isaac's dreams for us were contagious as airborne viruses, and we couldn't help talking from time to time about when we would be living on the lake, I would have been happy—and I'm confident Loretta would have been happy too—just to see Isaac at ease in this world we had brought him to, see him pleased with his achievements in it and smiling as he looked it in its eye. Of course, I'd also like to have a view of water. But so would my friend Henry Ah Sing, who owns a Szechuan restaurant in Old Town and is far more likely to get such a view than I am.

Because three months, one week and two days ago, on a typically dark, windy December night in eastern Washington, I watched Isaac follow his own missed sky-hook shot with a brilliant rebound and slam dunk over the seven foot Nigerian center who plays for Washington State, turn to head back downcourt, come to a complete stop while everyone ran by him, raise his arms a few inches toward his breastbone, and crumple to the court as though he'd been shot. He hit the wood floor so hard that I could hear the sound of it over the wildly beating drum that

Cougar fans were using to whip themselves into a frenzy as they urged their team on. I swear I could feel it through my toes.

The team doctor rushed onto the court. He threw himself onto his knees and skidded to a stop beside Isaac, jerking open a bag as he reached toward my son's chest. Almost instantly, they were surrounded by members of the team, all of them too tall to see past.

The image that stayed before my eyes was of Isaac's utter stillness there. He lay sprawled across the top of the key, his face down and twisted slightly to the left in a position exactly like the one he always slept in as an infant. My first thought, odd and unbeckoned, was that I was glad Loretta's arthritis was acting up so that she had decided against coming to Pullman with me.

I got up to run onto the court, but was restrained by some people sitting nearby. Their collective grip on my arms and shoulders—part embrace, part shackles—felt as though it was cutting off my air supply. I shook myself free and ran onto the court. The videotapes shown over and over again on the news, especially the one on ESPN, shows my mouth opening and closing as though I were screaming, but nothing was coming out. To me, it looked like I was trying to breathe for Isaac.

Technically, he was already dead when he hit the floor. But I didn't need to be told that. As I watched him fall, I swear I glimpsed a faint spray like the sweat that comes off a boxer's face when you see a slow-motion film of him taking an uppercut to the jaw. I understood that this was the spirit rising out of Isaac, a soft blue incandescence, it seemed to me, the actual formation of an aura around his collapsing form. When I knelt beside him, Isaac's gigantic body seemed vacated and when I touched the center of his limp palm—his most ticklish spot—there was nothing.

Miraculously, it turned out that Marius DePino, the west coast's premier heart surgeon, was in the crowd. This was certainly the first and most profound of our blessings that night.

Dr. DePino had come from Seattle to tour the campus with his youngest son, who wanted to become a veterinarian. When they'd heard that Oregon would be playing at Washington State, they stayed to see the great Isaac Berg in person. Dr. DePino quickly made his way down from the stands and trotted onto the court, working through the circle around Isaac's form until he was beside my son. His hand brushed my shoulder gently, asking me to step aside.

Marius DePino brought my son back and kept him here, turning the visitors' locker room into an emergency room (while getting me to scribble a waiver of liability on the back of a Cougars-Ducks souvenir program) and working on my son's heart before my eyes. The inventor of the DePino Procedure for correcting mitral valve prolapse and of the DePino Technique for grafting veins in bypass surgeries, the author of two cardiology textbooks and a popular novel about the mystical bond between a heart transplant recipient and his donor's daughter, the man was both brilliant and bold. I don't remember anything after seeing DePino's hands begin moving toward Isaac's chest. The team doctor got to work on me while DePino worked on my son.

There was more surgery later, after he was stable, to correct Isaac's hypertrophic cardiomyopathy, a thickening of the inner wall of his heart's pumping chamber. Usually, this disease can only be discovered after its young victim has died a sudden death. This is why I'm supposed to regard Isaac as lucky. They put a small electronic defibrillator behind his stomach muscles to shock his heart back into rhythm whenever it goes haywire on him.

Isaac always wondered how to get closer to God. To me now, it feels like he is—the Lord's hand present in tiny heartshocks emanating from Isaac's belly. But what it feels like to me is not what it feels like to my son.

You see film clips on the news all the time. These very tall young men in the greatest physical condition imaginable, these invincible kids on the brink of vast fortunes, suddenly collapse on the court. One minute, they slam dunk and make the whole backboard shake or they swat a shot into the fourth row; the next minute, they're dead. Hank Gathers of Loyola Marymount, dead. Reggie Lewis of the Boston Celtics, dead. Marcus Camby of Massachusetts, prostrate on the court, out of action for a few weeks, then coming back to play and no doubt terrify his parents. It even happens to women athletes, always the long and lanky ones, basketball or volleyball stars, the ones with slender fingers curling under toward their wrists, great specimens, dead. Marfan's Syndrome, a hole in the heart, a faulty valve, a heart too large, a heart hiding its flaws until the sudden failure.

There were three hospital beds turned sideways and pushed together with their side-bars down to accommodate Isaac's body. His first coherent postoperative words were spoken in a raspy whisper five days after his heart had stopped.

"Dad, I'm sorry."

"And I'm overjoyed, Isaac. You're still with us." I let go of his hand and stroked the side of his face. "Now let me go outside and get your mother."

"Wait." He squeezed my arm. In the past, he could bruise me with such pressure but now his grip was weak, child-like. "It's all over. They made that clear yesterday.

No more basketball. No running. A very quiet life. What, I'm suddenly going to become a physicist? Everything we planned for is out the window."

"I never planned for any of that stuff."

He closed his eyes and seemed to drift off. I shifted my weight to get up and leave the room, but he squeezed my arm again. "The house, the car, the trips, all out the window. I can't believe it."

"It doesn't matter, Isaac. Look," I pointed to his right, "all that's out the window this morning is sunlight." I couldn't believe that these were his primary thoughts after all he'd been through, after all he'd lost for himself.

His hand fell back to the bed. He swallowed dryly. "What good is a seven and a half foot tall non-basketball player? I've never done anything else. Never even thought about it."

"Don't exaggerate."

He blinked and looked at me closely, as though seeing me there for the first time. "I don't think I am, Dad. This is serious, what happened to me."

"Well I think you're exaggerating. You're only seven feet five and three quarters. Now let me go get your mother. She'll want to hear your voice."

When I brought Loretta back into the room, Isaac was asleep again. She looked at him, then back at me, her face filled with questions and worry. I bent over the bed and stroked Isaac's brow.

"What?" he whispered, opening his eyes.

"Nothing. You were dreaming, I think."

He moved his eyes back and forth. "He was here, it wasn't a dream."

I started to shake my head, contradicting him, but Isaac grew more agitated. It worried me and I thought about going out to fetch a nurse, but Loretta held me in place.

"I think I may be getting nearer to God," Isaac whispered.

"He's gotten near enough."

"I don't think so." He reached vaguely in my direction and I handed him the cup of water. "Just the outskirts, where the light turned me around."

"You can remember that?"

He nodded. "And a sound, something like a windstorm in the darkness, but filtered through a long stand of trees. I don't know." He sipped and handed back the cup. "This is something I can tell people about."

"Sure. When you're ready, maybe we can set up some kind of speaking tour."

He blinked. "I think maybe this is what I'm supposed to do, you know? Maybe it's why I'm so tall, to be closer to Him than most people."

Loretta nodded. But I didn't get what Isaac meant. He was so tall, we'd learned, because of a combination of genes and a pituitary disorder.

"Maybe," I said. "But you need to rest. You need to heal for a while before you even think about what to do next."

"This has to have happened for a reason," he whispered, closing his eyes.

"You were born with a thick wall in your heart. The pump was bad, that's the reason. You should get some sleep, there's plenty of time to talk about what you'll do."

"Nothing happens by accident," Isaac whispered. "Do you think Marius DePino was there by coincidence?"

"Yes," I said. "We're lucky his son didn't want to be an electrical engineer or they might have been watching the Gonzaga game instead."

Isaac closed his eyes. A faint hiss came from the machinery beside the bed.

Out in the hallway, Loretta took my arm and gently led me toward the waiting room at the end of the hall. We plopped down together on the brown plastic couch, sighing as one, relieved to have him talking but, I believed, a little shaken at what he was saying.

Loretta patted my arm, always a bad sign, and announced, "Well, you handled that about as badly as a man could, I would think."

"I'm not even sure someone who's a Jew by adoption *can* become a rabbi."

"We're reformed Jews," Loretta said. "So almost anything is possible."

I looked out the waiting room window into a parking lot where cars all neatly fitted into their diagonal slots gleamed in winter light. It seemed possible to rearrange their pattern, to shift them so they faced the other way, noses to the street instead of noses to the hospital wall, just by fluttering my eyes or twitching my pinkie.

"For some reason, I just don't like it. The idea makes me squirm."

"Who says you have to like it?" Loretta asked. Her tone wasn't particularly challenging, just curious, a gentle questioning. "I never much liked the idea of his being an athlete, myself."

"You're kidding me."

She shook her head. "So I didn't come to very many of his games. Big deal. You could skip his services, or start going to the Conservative temple, whatever." She paused; I could hear her swallow. "The point is, George, you don't have to like what Isaac does."

I nodded, but a person would have to be looking closely to tell. Of course she was right, I should just be glad he was still here to do anything. "Tell me something." I paused, still looking down at the cars, picking one long Lincoln and changing the angle of my head so that the reflected sun pinged off the fender like a shot. I got lost in the game of it.

"All right. But first you have to ask it."

I turned to face her. "Why do you think I'm uncomfortable with this?"

"Beats me." She got up and came over to me, putting her arms around me from behind and hugging me to her. "No it doesn't. You're terrified, my love. You're in shock. You want everything to go back to where it was two weeks ago."

I nodded, leaning back into her. "Probably right."

"And you've had enough God for a while, I think. You don't want Isaac inviting Him back into our lives any time soon."

"What if it's something else?"

"Well, then it's something else. Now why don't we go for a walk and let him sleep."

"In a minute."

"Ok, George, what is it?"

"The new house, the recliner, the trip to Rio?"

"You hate the beach."

When we brought Isaac home two weeks later, the house immediately seemed tawdry to me. It felt too small to contain him, though he'd lived there with us all his life, and our furnishings looked shabby in a way I'd never seen before.

I've worked hard for thirty-one years now, and this is all I have to show for it? Three decades selling people insurance, eventually opening my own office, making nice money, and I live in a house I'm secretly embarrassed to own? We haven't had friends over to dinner in years—I always thought it was because we were too busy, too involved with following Isaac's career or with our little projects, our routines. Now I realized it was out of shame; who wants to bring people in here, with the faded wallpaper and dull paint and threadbare carpet? It was as if we had stopped tending to the place five years ago, almost exactly when Isaac's potential as a basketball player revealed itself. Everyplace there was wood there were chipped surfaces, as though the house had been subject to airborne abrasion. Posters of flowers and vegetables were wavy and sagging behind their glass, windows whose seals had busted were all blurry with contained moisture, there was noise everywhere from appliances that labored to keep up. All of a sudden, I could see the place for what it was, for what it had become. An abandoned home.

While Loretta tended to Isaac upstairs, I found myself sitting in my old easy chair staring out the window. The view, what I could see of it, was east toward a commercial section of Portland. A flickering neon sign with two letters out told people driving on the below-grade freeway that home furnishings were for sale. CAREYS FURN URE. Across the freeway, a new medical center loomed. The day seemed unnaturally dark. I turned back and listened for the sounds of my wife and son, his deep murmuring voice, her high breathy melody of comfort and devotion.

The doctors had told us that Isaac had a very good chance for a satisfactory outcome, whatever that might mean. His heart was essentially good, which I could have told them without holding it in my hands, though slightly enlarged by years of extra pumping action. His circulation and respiration were all good, he was in great condition, under the circumstances.

It was just that he couldn't play basketball again, or engage in strenuous activity, or do much of anything he'd always done and always dreamed of doing. Perhaps what bothered me more than all this was Isaac's reaction. He seemed happy, relieved. He was still tormented by the feeling that he had let us down, broken his promises, but with help he was getting beyond that. He seemed, in fact, gradually to have grown joyous about the new direction his life would take.

What, did I want him to be depressed? To be immobilized by despair over his losses, to be so angry that he jeopardized his recovery? No, of course not; but giddy, as though he'd been let off the hook? It was all very confusing to me.

We had made inquiries about how Isaac could begin studying for the rabbinate because that's what he asked us to do. I'm sorry to admit that if Loretta hadn't taken over, it might not have gotten done. Neither of us—nor Isaac—had realized how many credits in philosophy and religion he'd already accumulated. Still, it would be a good five more years, provided his recovery continued apace, before he was likely to stand before a congregation of his own.

I turned back again toward the window and began to drowse. A glimmer of color erupted outside the window. At first I thought it was just the Carey Furniture sign blinking, but then it came back, a small bird darting quicker than any point guard I'd seen on the basketball court. It was the most outrageously bright yellow, with hints of black on the wings and tail, a brilliant red head, and it zoomed across my field of vision like a flash of sunlight bright enough to cut through the barrier of the cloudy window. I thought I could even hear its hoarse call, its *pit-ik pit-ik* over the traffic sounds and the voice of the refrigerator cycling on in the kitchen. What's a western tanager doing a half-mile from the Banfield Freeway, I thought. What's it doing with a red head in winter?

Then, jerking up with a start, I realized that was where I really wished we were, the three of us, where the bird was. We should be in a small house in the woods, in springtime. One of those yurts maybe that come pre-designed so you can assemble them yourself in the middle of your acreage. Not by an urban lake in a big house paid for by my son, but a cozy little place without many walls, located halfway between the city and the shore, precisely the kind of home I'd always dreamed of having. I had five years left for selling insurance before I was ready to retire. By then, Isaac would be finished with his studies. Could he find a congregation for himself in rural Oregon?

I stood and went over to the window. Nothing. At least, nothing in the way of yellow birds. Still, it would be nice, I thought, and turned to head upstairs to see what Loretta and Isaac might think about my dream.

Judy Johnson-Williams, *Not Finding Treasures* (detail), 2000

Riders on the Earth

Valerie Hurley

Shortly before my second birthday, the atomic bombs were dropped on Hiroshima and Nagasaki. My mother arranged a birthday party for me, pinned a white silk rose in my hair, dressed me up in taffeta and Mary Janes, adorned the dining room with favors and paper hats. Hiroshima and Nagasaki were filled with Japanese children. My mother took pictures of my face stuffed with cake, the flower dangling down by my ear. Eight two-year-olds sat on the sunlit front step, squinting into the camera.

My mother believed that the United States had to drop atomic bombs on Japan. She believed that breast milk was bad for babies and that she needed the silvery blue DDT bomb under the sink to grow the zinnias and marigolds in her garden.

I was nurtured by a world unsure of itself. The neighbors constructed fallout shelters. In school there were air raid drills, and we were taught to crouch down in the hallways. What was about to come shrieking through the glass? Who would bomb our school? Would our arms over our heads preserve us?

In 1962, I sat in the college lounge listening to Kennedy's ultimatum to the Soviet Union during the Cuban missile crisis. It suddenly seemed obvious what the years of squatting on the cold tile in the school hall had been for. Terrified, I telephoned my father in New York. He laughed at my alarm. I was comforted by his nonchalance.

The fifties were not a time when things were talked about. Menstruation maybe, but not sex, and not war. The fifties were a time of atmospheric testing of nuclear bombs, strontium-90 in the cows' milk, DDT on the crops, fallout shelters in backyards stocked with a year's supply of pickles and dried beef.

My neighbor went to college in the mid-fifties, and it took three bedrooms of her house to contain her new wardrobe. I was one of the neighborhood children who toured the rooms, dazzled by the strapless net gowns and gloves and plaid pleated skirts and cardigans and matching purses and shoes. But eight years later when I went to college, what was needed was not a trousseau but some blue jeans and a moral vision. It was the time of the Beatles, Vietnam. The toddlers of the forties now wondered more and believed less. We were questioning some of the things we had been taught—respect for our elders, obedience to authority.

In 1969, my new husband and I lived in the highest-crime area of Manhattan, in two long narrow rooms decorated with paper lanterns from Azuma. There we were mugged by two jittery young druggies carrying knives. Eventually we moved to the country, looking for a purer world. I envisioned brown eggs and strawberries, spring water, an orchard, a farmhouse with a fireplace in every room. Instead, we got mice nesting in the box spring, a leaky roof, a flooded cellar, young fruit trees nibbled by rabbits, deer, and caterpillars, red squirrels living in the attic. We were happy. Here, for the first time in twenty-seven years, we were living close to the Earth.

The summer before our first daughter, Mara, was born, I planted a tulip garden for her. She was born in February, and in May big, ruffled tulips appeared—pink and yellow and red and scarlet and ivory. I walked her carriage down the road, tucked bunches of wildflowers in beside her face, hoping she'd grow up to love them.

When our second daughter was born, they wheeled her back into the room after the delivery. Erin was less than an hour old, but she looked up at my husband and me and studied us for a long time. Her eyes were wide and alert, large and blue; it was not a loving look. *Who are you?* She seemed to be saying. *Are you worthy of me?*

I nursed and rocked my babies, sang to them, introduced them to the flowers that grew around our farm, the moles that ate the tulips, the rabbits that ate the pear trees, the deer that ate the apples. My husband created games for them: Push the Daddy Over, Daddy Elevator. They shrieked with delight. At night we'd often all crawl into our big bed and "chitter-chatter" in the dark. We gardened and ran in the fields; we sat on a boulder by the brook and presented each other with pebbles wrapped in maple leaves; we picnicked and ice-skated and lay in the grass beneath the sugar maples.

I read the children the story of a bald eagle that was hatched under artificial conditions. They sat in wonderment, their arms hugging their knees, their soft hair blowing in the breeze. Because of the buildup of DDT and other chlorinated hydrocarbons in the bald eagle, the eggshell has become so fragile that it often breaks when the female eagle sits on it. So in an early stage, biologists retrieve the delicate egg from the eagle's nest, perched high in a tree or cliff, and place a plastic facsimile in the nest, which the mother eagle continues to sit on. The real egg is transported to a laboratory to incubate and eventually is returned to the nest. The parent eagles accept the exchange, and the eaglets hatch and survive. We looked at the photograph of the scrawny new eaglet—a creation of its time.

Stretching out around our yellow farmhouse were fields of wild mustard, buttercups, clover, yarrow, butter-and-eggs, slowly withering into goldenrod and chicory, Queen Anne's lace and lavender asters, wild morning glories coated with frost.

On the last day of March, the peeper frogs started to sing, and they sang all spring, all night. In April, farmers tore up patches of soil with their plows, and I dug with my shovel, easing in frail fruit trees, denuding the few branches and then stopping for a moment to touch their spindly trunks and wish them luck. The full moon rose above our house. Peach-colored clouds floated by at twilight. A dream was in the air, the dream of continuity: the Earth's light and shadow, its color and monochrome, its stillness and change, its fragility and concord.

Beyond the lavender mountains, eight thousand nuclear bombs had been planted in the wheat fields of the Midwest, encased in huge Minuteman silos. Some days, I could feel their presence. Trident submarines crept through golden forests of kelp. My father would accept their presence without comment, as would my mother. They lived in a gentler time of crisp Hollywood endings. But I was living in an age when missiles were labeled Peacekeepers and when well-intentioned biologists made the precarious journey up into a tree to redress the harm we have done to the thin polluted eggs of our national bird.

In a teepee at the edge of a stream, I camped with my daughters one night in August. The teepee was constructed of a canvas cover stretched over seventeen stripped tree trunks. We lit a fire inside the teepee and roasted corn and potatoes, watching the smoke rise up through the smoke flap, gently blurring out the stars. After dinner, we walked in the dark, combing the earth with our flashlight. Goldenrod bent to touch us as we passed; in the tall grass, the cold lights of fireflies gleamed. The children squealed with fear and delight. Returning along the path of larches, we saw the teepee standing before us like a goddess, a huge white cone blown with shadows, swaying with fire breath.

I tucked the girls into their sleeping bags and lit two candles. I tried to read for a while, but the teepee was not a space for words. On the flapping white walls, my shadow wavered. Chunks of potato sizzled in the embers. The children slept as I kept vigil, watching their pink cheeks, their warm hands, the dark curl of their eyelashes.

In the middle of our bedroom floor, my daughter Mara sat one night with a large colored globe of the world in her arms. She studied it, frowning, for a long time, then looked up at me and asked, "Are we trapped on Earth?" The expression on her face told me that she knew we were and that this was not necessarily a good thing. I often think of her in that pose, with her brown-gold curls and her long legs, her blue eyes filled with perspicacity as she sat on the floor with our small round Earth pressed in her arms.

When he was twenty-four years old, the astronaut Russell Schweickart was a fighter pilot stationed in the Philippines. Every fourth week it was his assignment to stand nuclear alert at an airbase on Taiwan. Four planes were parked at the end of the runway, fully fueled, a nuclear weapon slung underneath each. He later wrote, "After much agony, I came to realize, knowing what I know now, that if I had to decide again, lying there under the stars, my back pressed against the bomb, I wouldn't drop it. . . . I now understand that we can't pass along such decisions to higher authority, for there is no higher authority than that which exists in each of us, individually, as we face our complicated and ambiguous world. . . .

Our dilemma becomes increasingly daunting. Will our vision of the human future be large and clear enough to lift us beyond the uncertainties and fears of our cosmic birth? Will we have the wisdom and courage to accept the individual moral authority within each of us? Or will we defer to experts and impersonal systems of authority in the false belief that in them reside greater wisdom and morality? In how we answer these questions may lie the outcome of the great experiment of life."[1]

When Mara was eleven, a tall thin girl with curly hair, a lively mind, and a joyful spirit, she was rushed off to Boston for brain surgery. Before leaving, she wrote a note and put it on the table: *Mara Was Here.* Seeing it, I pictured us returning from Boston without her and finding her note on the kitchen table. "What did you do that for?" I snapped, tearing the paper to pieces.

It was the shoot-out at high noon with God. I never felt able to bow to the mighty will that was thrusting itself against mine. I didn't care if it was God's will, or destiny's—I wasn't about to submit to it if it meant that this child I so loved would be snatched away. I would not allow it. I refused to sit by passively, a good girl of the fifties.

After two surgeries, radiation treatments, brain infusions, chemotherapy, and nine months back in school, she became very ill. She couldn't walk. She hallucinated. She was different now: passive, confused, sleepy.

The brain tumor had spread, and the doctors gave up hope for her. They described her cancer as idiopathic—of unknown cause. Some of the people in my family, trained in the fifties, the age of trust, the age of plenty, the age of optimism, the age when you boldly fortified yourself against your enemies, called to say that everything would be fine. They recounted miraculous brain tumor stories. But would the prince really gallop in to rescue the shining child? I thought of the shining children in Hiroshima and Nagasaki. My family reassured me, but I was the under-shadow of their confidence. What I had absorbed from my childhood was not the might, the brave front, the war whoops, but the bluffing, the fear and mistrust, the pretense.

My husband left his job and learned the nursing skills required to care for our child and, together with our eight-year-old daughter, we kept her in a hospital bed in our living room for the last five months of her life. I staged every form of ritual: prayers, healing masses, baptism. I dabbled in Christian Science. I put her on diets, telephoned experts on healing. Their advice was just to *be*. Allow my dying child to drift farther and farther out of my reach? I kept her in our lair: sang to her, piled her with blankets, lit crackling fires in the woodstove, tried to create a womb that she, passive though she was, would never leave.

In the living room where she lay, there were six windows. The leaves fell and stuck to the glass, lilacs on the high bushes changed from lavender to brown, snow fell and melted and dripped down the windowpanes, then turned to gray sleet. I wanted to exclude nature from the room and went in several times to measure the windows for curtains. One day in October when I had been measuring the windows again, Mara said, "What do you think, Mom, the windows are growing?" I never did get curtains. They could pretend, they could lie; they could flutter prettily in the chill autumn breezes, but they couldn't hide the conspiracy of nature that colluded with greater force each day just beyond the windows. I refused not to look. I would sit. I would sneer. I would never submit. I would be like Winston Churchill who said during the Second World War, *We shall defend our island, whatever the cost may be, we shall fight on the beaches, we shall fight on the landing grounds, we shall fight in the fields and in the streets, we shall fight in the hills; we shall never surrender.*

In spite of the immensity of my will and the length of our vigil, one sunny January day our beautiful Mara died.

For two years after that, I tried to carve a home out of the ruins. But the land itself had been altered for me. Nature could not retain its innocence. It was nature that had dragged her away from me—the trees and the grass and the clouds as much as the trickster disease or the bony arms of the undertaker.

As I watched her die, this land, once beautiful, and all the contiguous land, rearranged itself according to its witness. It absorbed the cruel scenes it had observed. Cast upon the high, swaying lilacs is the image of Mara lying, so thin, in the summer grass. Frozen into the landscape is the cold January night sparkling with frost when she left home for the last time. Nature appeared to be indifferent but in fact was not. The stars absorbed the sight of her being carried away; into their glitter the scene is woven tight.

One summer day I went up to her grave and planted a chrysanthemum plant. It had many bright pink flowers with yellow centers. Afterward I lay in the grass in the sun for a long time, beside the plant. When I opened my eyes, I noticed—in my green pants and bright pink shirt—how well I blended in with her grave. It was time to look for a new home.

I would turn my back on the cemetery I hated so, the school bus that passed the house without hesitating, the junior high school she never attended, the drugstore where I rented her wheelchair—all the places where she should have been but wasn't. We had many friends, but friendship, love, warmth, however strong and comforting, could not correct this realignment of the earth.

Vermont called to us, with its chant of the pure North. Scarlet apples hung from the trees. The cornfields were bleached gold. Alfalfa curled over the bony hills. We moved into a rented house on Lake Champlain, facing the purple Adirondacks. It was made of brick and reminded me of the three pigs, and the house the wolf couldn't blow down. It was comforting, living in this place, with its accoutrements I would not have chosen: wall-to-wall carpeting, microwave oven, basketball hoop, brocade chairs, linen dresser cloths. In a dark mahogany bed that has history in its bones, I lay and thought of the people who had slept in that bed over the decades, their joys, their worries, their griefs. I wondered what makes a home, what destroys it. It seemed to have something to do with what's kept out and what intrudes.

In Vermont, I visited a young man, Martin Holladay, who had been imprisoned for breaking into a nuclear facility in Missouri, hammering the silo cap and painting the message *No!* on the cement. Martin was raised in Lebanon and, after some time at Yale, had become a farmer. He lived without electricity in a cabin he built in the mountains of northern Vermont. One day, he sold off his animals, closed up his little farm, and hitchhiked across America to climb the fence of a nuclear installation.

Martin writes of his experience: "Part of the reason of our profound failure to deal with these nuclear weapons on a moral level is that it takes an act of the imagination to understand the reality of our huge arsenal. The traveler sees only a fenced level area marked with a 'No Trespassing' sign. But the reality of that site is a Minuteman II missile with a range of eight thousand miles, armed with a 1.2 megaton nuclear warhead, one hundred times more powerful than the Hiroshima bomb. . . . For each silo the earth has been excavated and replaced with concrete, steel and plutonium. The missile is in the cornfield: our separation from the fields is now triumphant. . . . The insertion of a forty-foot nuclear missile into a buried silo is a graphic image of rape. We are sowing a different crop now, and none can imagine the harvest."[2]

I think of Martin's courage, and of the courage of my husband and daughters, as I read Archibald MacLeish's words,

> To see the earth as it truly is, small and blue and beautiful in that eternal silence where it floats, is to see ourselves as riders on the earth together, brothers and sisters on that bright loveliness in the eternal cold.[3]

Notes

1. Russell Schweickart, "Our Backs Against the Bomb," *Discover* (July 1987): 62–65.
2. Martin Holladay, "Journey to Missouri," in *Swords into Plowshares,* ed. Arthur J. Laffin and Anne Montgomery (San Francisco: Harper & Row, 1987), 143–144.
3. Archibald MacLeish, "Riders on Earth Together, Brothers in Eternal Cold," *The New York Times* (December 25, 1968): 1.

III

Places in Time

Childhood in the Church of Darwin

Leslie Van Gelder

I was raised in the church of Darwin. My father was an evolutionary mammalogist with a predilection toward skunks, bats, and African antelopes. Our home "parish" was the American Museum of Natural History, where my father was curator. During his thirty years at the museum, he studied bats off Baja, built a houseboat and sailed down the Amazon in search of mammals, and spent the last twenty years of his life taking annual pilgrimages to East Africa to study nyalas, kudus, and human tourists. In his museum cathedral, he designed the blue whale that hangs gracefully over the Hall of Ocean Life, the very symbol of the American Museum of Natural History. He always told me she was my sister.

What incense, oil, and stained glass windows are to some, formaldehyde, study skins, and glass cases are to me. While my friends watched Charlton Heston's *Ten Commandments* every year on Easter, we watched the PBS miniseries on the *Voyage of the Beagle* with the same sense that we were witnessing a holy story coming to life. In between rebroadcasts of Darwin-related miniseries, we watched National Geographic specials that dealt specifically with mammals. My father had the capacity to identify lions as if old friends—often correcting the announcer by claiming, "That's the Okavango, not Ngogoro. She's not the same lion they showed a minute ago, even though they're claiming she is!" He'd usually end with a loud, "Jesus Christ," associating this stranger with the capacity for stupidity, not a sense of wonder or love.

We traveled the world with my father—from his beloved Zinave in Mozambique to Arizona's Sonoran Desert, to the high ranges of the Cascades. From him, we learned lessons in animal identification, ecosystems, and whole animal biology. But what he never spoke of was spirituality.

I've often wondered why, since it is clear that what drove my father's research was not the hours of repetitive note taking or trap setting, but the moments of wonder. Writing to a friend from his research camp along the Savuti Channel in Botswana in 1976, he attempted to capture his love of place. In a letter, he writes:

> This probably sounds absolutely mad to you, but there is something quite Eden-like about it. You have all the real needs of civilization—shelter, food, transportation—and all of it comfortable, but you are plunked down in the middle of virtually unchanging wilds, where at any time you see thousands of zebra and wildebeest, giraffes and elephant, and the hippos in the river. Overhead there are birds by the thousands, and at times of the year by the millions, so that they look like smoke when they stream from the trees, and their weight actually breaks the branches. Lions and hyenas abound, and the whole basic cycle of life and death takes place before you.[1]

When I took to traveling on my own in the 1980s with a first foray into teaching environmental education in Newfoundland, my father and I spoke of our parallel experiences. The love of coming to know a place—the intimate details of understanding how to read the weather or learning the local shortcuts—knowing the names of plants, birds, and the town bartender. For each story I told, he gave me one back. We spoke a language of resonance. We never articulated the driving force behind both of our pursuits: our love of place. Connecting with the natural world, we found something in ourselves that we never gave a name.

Perhaps this spiritual connection was a given, but I think we never spoke of spirituality because the language of religion was too tainted. Growing up, Hebrew school and the "Temple Building Fund" became synonymous with the financial aspects of organized religion. We sensed a shallow commitment by the leaders to the word and the practice that underlies the faith. The rabbi didn't show for my bas mitzvah because it was snowing. *We* had to be there, but he didn't want to dig his car out of the driveway, so the tenth man for the minion was a drunk from a bar across the street. This didn't do much to impress on me the level of commitment inherent in our synagogue. The boy who was bar mitzvahed with my oldest brother chose to ride into Temple Beth El on an elephant since his family had just donated a new wing. Perhaps it was during this time that I developed a degree of doubt.

Spirituality got entangled in the language of organized religion, and the only religion we could espouse openly was our belief in the church of Darwin. We spoke of Darwin as if he were our most revered ancestor. The Galapagos Islands took on the sheen of Mecca. Someday we would make pilgrimages there. My brothers and I each owned our own dog-eared copies of *The Origin of Species,* with annotations in the margins to support our research interests. Mine lent itself to a thesis on Darwin's influence on nineteenth-century children's literature. My eldest brother conducted genetic experiments with drosophilae. My other brother named our cat Darwin and his supporting hamsters Wilberforce and Huxley. Huxley outlived Wilberforce by two years. Darwin is now pushing twenty and has outlasted both of my parents.

The human Darwin was more than merely academic fodder; he became something of a kindred spirit in our explorations. As my father had spent his first years as a mammalogist doing research in Uruguay, Bolivia, and Argentina, the stories of Darwin's journeys resonated with his own nights out on the pampas among the gauchos. Closer to home, when my brothers suffered terrible seasickness on a boat trip across Lake Superior, they were comforted with, "Darwin was seasick for the first few months he was on the *Beagle.*" Their own bit of seasickness connected them in spirit to Darwin. Further, to convince my mother that the pet snails I brought back from Maine were to be permitted residence in our house, my father mentioned that "as a child, Darwin was notorious for collecting snails, beetles, and small birds." The snails stayed. No one could argue with the weight of Darwin.

My father's own childhood was filled with the wonder of the natural world, even the natural world of Manhattan in the 1930s. According to family lore, squirrels were raised in the sashes of curtains. Dead insects, plants, and rodents were subject to storage and study, and the Museum of Natural History and the Bronx Zoo became playgrounds for the imagination. The feeling of wonder spurred the explorations, and his studies confirmed the sense of wonder.

My father was a natural historian in the Victorian sense. A collector. What his friend, the writer Peter Matthiessen, once jokingly termed a dinosaur—one of those old guys who knew not a little but a lot about a lot of things. The sort of generalist who was acceptable in a world where people believed it was possible to see the connection of parts. Today we are much more familiar with the specialist who has deep niche knowledge but is rarely global in perspective. My father took degrees in zoology and literature, studying medieval depictions of animals alongside academic journals of mammalogy.

Like Darwin, he was a natural taxonomist—a believer in categories, not because of their inherent value, but because he believed there was a bigger picture. Darwin saw the natural world as a unity, subject to a series of natural laws. For my father, and perhaps Darwin, those were the seeds of spirituality. It came from a sense of faith in a world where systems were at work. With each verification of the system, be it evolution or nonlocality, faith deepened—faith in a world with a logic of its own. If we could come to understand its mechanisms, we could find a sense of order and meaning. Wonder fueled questions. Questions fueled research. Research fueled a desire to verify that the feeling of wonder was indeed part of a greater process.

To understand the processes of the world at work, I took a track different from my father. While skunks and bats have their charms, I was much more interested in the way some cultures had developed a deep understanding of the natural world. Cultures with local knowledge. Cultures where there is an active awareness of the process of nature as more dominant than the process of human manipulation. My interest arose out of my experience living in the southern Labrador straits, a place where the natural world dominates people's lives.

I lived on an island with a population of 300. Many were second-generation Canadian, their ancestors English and Irish fishermen. Most dropped their "H's" so the town of Harrington Harbour, was 'arrington 'arbour, but they found their "H" in order to get us h'eggs and h'ice. In town, the mayor was also the mailman and the garbage collector. The boat that served as bank, food delivery service, and the only "highway" for hundreds of miles came through once a week. Time moved differently there than it did at my home, Harrington Park, New Jersey, ten miles from Manhattan.

For the first month I was in Labrador, I rushed about on a self-induced schedule. "Why wasn't the mail being delivered today?" I demanded. Because it's foggy and the plane can't land. Why were we trapped in the seabird sanctuary for five extra days, forced to eat macaroni and cheese and corned beef hash, at every meal? Because the winds were too high for a boat to get through. In a place where the powers of nature dominated, my attempts to tame time took a subordinate role to nature's plans. Each time I complained about the weather getting in the way of "my plans," I was met with the tolerant humor adults often give to their children. Each "why" also prompted the gift of a story.

It could have been the story of a shipwreck when someone was foolhardy enough to go out in a storm. Sometimes it was the story of the worst winter snow they ever had when they were sure they would starve. Or the story of the nurse from New York who was killed by a pack of sled dogs. Once I heard the story of a man they believed could tame the northern lights by playing his fiddle for them. These stories, I realized, were a way of teaching me how to cope with the vagaries of life on the Labrador shore. All stories came with the caveat that if I planned on living a "less difficult life" there, I had to learn that the natural world was going to do as it pleased and I, not it, would have to adapt. In a place known affectionately as "The Land God Gave to Cain," I could live successfully if I learned nature's ways. Local knowledge.

They were right. Once I did slow down, I understood much more. I could spend an afternoon watching ravens fly. Imagine, a whole afternoon, lying on a rock face watching ravens. Sounds like a waste of time from many points of view, unless of course I could claim I was a scientist officially studying ravens. I wasn't. I was simply learning.

After an afternoon on the rocks, I knew how ravens flew, knew that they curled their wings around the edge of the wind to control their ups and downs. I knew that crows made too much noise for me, but ravens were mostly silent, moving in an orchestrated ballet. I could tell ravens from far off not by their size or shape, but by their movement. I learned why one area was called Crow Head and another point, where the ravens and gulls flew, was not: ravens love updrafts; crows do not. This was local knowledge. It came from living in a place.

The emphasis on place is common among people who have lived in one area over many generations or in locations where humans do not get much foothold or control, like Labrador. Place-oriented cultures develop a deep-seated respect for their environment and tend to measure wealth in terms of relationships, using respect as the system of currency. Beings become animate, storied entities worthy of respect and possessed of an individual wisdom. Time becomes tied to place as humans watch the seasons pass, measuring change from year to year through the familiar transformations.

In our rooted sense of being, we are all individual places who maintain a singular point of view. Every person is a place. It is perhaps easier to imagine that the first place we ever were in the world was also a person. Our mother. Imagine sending mail to each other—"Fred Jones c/o Marcia Jones Uterus." We moved separately from her but could not move outside her. We were in our own amniotic sac but also connected. From the very beginning, we experienced a unique environment that no one but us has shared. We were in a constant state of change, adding new cells, organs, sexual parts, developing a lot of fur and then losing it, until one day when we packed our bags and left the old neighborhood.

If doctors were to conduct exit interviews with newborn babies, they would probably hear climate descriptions or stories of favorite kicks. Perhaps descriptions of a great love for garlic or an equal hatred of marmite. Doctors might hear of frustrations over the sounds of a dog barking or the love of the rhythm of Mom's walking. From the very beginning of life, our experience of our environment helped us develop as unique individuals. At birth, we are not factory-produced tabula rasas; we are already experienced beings with stories to tell (though perhaps not the ability to articulate them).

From birth, we move into a different world. Our place is no longer our mother's body, but now we are attached to another form. Just as the umbilical cord once tied us to our mothers, we now have lungs and breath, offering us a relationship with all other living beings. We take air into our bodies, exhaling out a different version. In this way, we continuously connect to the physical world. Through our mouths, we take in food, and this too allows us to form connection to our environment. For as long as we are alive, we are in direct contact with the world, movement, breath, and nourishment. We are here.

We discover that we are part of an interesting continuum. Although we are an individual, in the sense of our mobility and our thoughts, we are also connected to all other creatures that breathe, eat, or move. Being a unique body and yet an integral part of something larger forms one of the great paradoxes of our lives. How do we feel both individual and special and also acknowledge our connection to all other beings?

One way to look at that question is to consider ourselves as places, as geographical points. Imagine that the world is an enormous space of relations. Each being is moving along through time, experiencing the world at every moment (some more consciously than others). In order to understand our experiences, we create a narrative that is very much our own. We become, literally, points of view; the result of environment + experience over a period of time. Since the world is a continuous process, each moment adds new experiences to the last.

With this in mind, we are always at a particular place in our life journey. Our uniqueness comes from our experiences and the ways in which we choose to interpret them. *Who* you are is literally *where* you are, the story you choose to tell. No one but you has traveled your route. No one but you has seen out of your eyes or felt the joys, pains, and sorrows that make up your internal landscape. This is

Thoreau's miracle—"Could a greater miracle take place than for us to look through each other's eyes for an instant?"[2] Because we cannot, because our journeys are our own in the sense that no one else has traveled them, we must find other ways to communicate what we have seen. We bring our stories wherever we go, and when we tell them, they illuminate what physicist F. David Peat calls the process of "coming to knowing." Peat says:

> Knowledge in the traditional world is not a dead collection of facts. It is alive, has spirit, and dwells in specific places. Traditional knowledge comes about through watching and listening, not in the passive way that schools demand, but through direct experience of songs and ceremonies, through the activities of hunting and daily life, from trees and animals, and in dreams and visions. Coming-to-knowing means entering into a relationship with the spirits of knowledge, with plants and animals, with beings that animate dreams and visions, and with the spirit of the people.[3]

However, if each person is the collection of a growing number of stories, how can one set of stories be superior to another? In a world of commodity, those who have more (whether money or college degrees) take a higher position over those who have less. Further, whole corporations can dictate what is "good" or "bad." In a place-oriented culture, however, each person's experiences must be valued toward a greater good, and none can appear superior. Instead of having a system of hierarchy, life is like the strands of a web, where the individual nature of each creates the collective strength of the whole. Passing stories from one to another makes life and survival possible.

In cultures that value place, respect for the individual is paramount because each individual is in effect responsible for other beings, learning from the other's experiences instead of imposing one overarching point of view. But how do we develop relationships with other places, whether they are people or geographical points? The individual is a place but also part of something larger, like a child inside its mother, a constantly changing, growing entity.

So, then, where are we? For the answer, we have to challenge the notion that all of our locations are physical addresses. Instead, we have to imagine a collection of processes within processes, continually moving.

I have two addresses. One is in New Jersey and the other in Oxford. When I have mail sent, I know that, barring days when the mailman has decided that he is afraid of my barking dog or Kitty the pernicious guard cat, it will be delivered to my home. To the post office, I am reliably unchanging. On the other hand, in the United States, I live in a house that has been evolving for the past thirty-two years. Yes, the quarter-acre is recognizable in the sense that it has not moved much, but I have seen trees sprout up and come down, seen the edges of the property encroach upon the next, and seen the house in which I live switch ecosystems.

Places have the power to show us our own process of change. If we return to the same place at different points in our lives, we begin developing a continuous narrative structure from the experience. Using location, we come to reflect on the ways in which we have come to know ourselves. Repeated encounters with certain places offer us three important tools to understanding the world's natural process. First, we have an opportunity to develop animistic, live relationships with our environment. Second, we develop a narrative framework with clear points of contact from first experience to the next. Third, we have direct conscious access to the process of time.

How does it work?

We may begin a story with an experience in a location. Our encounter with that place burns into memory and becomes the seed of a narrative. If we encounter that place a second time, the story grows. The return creates the replaced memory. The story now becomes two stories: the first one, seen through the past, and the second, connecting that vision to the present. Standing in place, it is possible to encounter both points of view. Memory gives one vision, the present provides another. In the space of a replaced memory, we have the possibility of moving through time while being rooted to place. We are as Senex, the mature farandole, says in Madeline L'Engle's *A Wind in the Door,* "fully rooted and able to move."

Being conscious of our replaced memories allows us to weave our stories together. We absorb our environment in those moments, knowing the feel of the land around us, allowing it to take root. This need not be a natural environment. One can just as easily replace memory in the stairwell of an apartment building as in Serengeti.

Western culture has generally purported to believe in a linear trajectory of time. Events happen in order. History is drawn on a time line with clearly delineated dates. One thing happens before another, and there is no going back—only movement forward. Time becomes a discreet unit, one that may be broken into pieces, used as a commodity for measurement and control.

Indigenous people and contemporary physicists tell us that time does not function as a linear arrow of continuous progression. According to the theory of relativity, space and time form a single fabric that can bend, shift, speed up, or slow down. Cultures that value place and develop reciprocal relationships with their environments tend to view time as closely tied to space. Time spirals over place, functioning like a coiled basket or tree ring, each year adding another layer. One moves in multiple forms of time, as past moves into present through the triggers of place and memory. Naturalist Terry Tempest Williams writes, "It is also a Navajo belief that if something happened once, it may happen again. In fact, if something happened once, it is likely to happen again and maybe again and again. The Navajo faith in the cyclic nature of things has come to them through their direct interaction with their physical environment."[4]

Western culture prefers to spend time talking about the future as being ahead of us. Taoism, a system of beliefs built largely on the study of nature and change, draws a very different picture of time, one in which time is tied to stories.

Let me paint for you the Taoist picture of time. Time is a river. We humans sit on the rocks on the side, facing downstream. Our future rushes up from behind us unseen. Our present glides by us on the rocks; we are unaware of its continuous presence. Instead, we look on our past as it unfolds in front of us. In the pools and eddies, rapids and flat-water, we have a fluid, connected motion to which we must always assimilate the new into the old.[5] Into this flows the places we have been and our memories.

Places are the structural elements of our stories, the rocks to which we can attach weight, the banks of the river that shape the turns we take. Some places are revisited as our lives circle around particular stones; others are rushed over, never to be seen again. Time, dreams, and memory become the tributaries that feed our stories, as we require opportunities for connection and reflection to make sense of the continual wash of experience. We need the continuous fusion of all three—place, experience, and storied reflection—to exist. In place, we find connection. In experience, we live. In story, we find meaning.

To be at home in the world is all of those things. The landscapes in which we find ourselves at home are our stories. To feel at home in the world means bridging the gap between your sense of being a lone individual and feeling connected to the processes at work. Your experience then becomes part of a greater story—what is often referred to among Native Americans as *Mitakuye Oyasin,* "All My Relations."

My father, the mammalogist, had faith in the process of evolution that came from testing ideas and observations against the processes he observed. My father's wonder in the world came as much from learning the rules of the external environment as it did from finding those same rules mirrored within himself. Charles Darwin, his hero, was an agnostic admitting, with a degree of humility that my father shared, that we just don't know. Why spend energy trying to prove an "out there"? Instead, grounded on earth, ask how we can live harmoniously here and now. How do we find a sense of faith in our present environment? Where is our wonder? If we love our home and love those within it, will we care for it as we do our children?

Recently, I brought my questions to Oxford's cathedral of science, the University Museum of Natural History. There, beneath the arches of polished English stone, I looked at the statues of Darwin and Newton, Linnaeus and Bacon. The saints of the "church of science" brought me no joy. Why worship them? They would exhort us, I believe, to turn our energies into discovering our world instead of contemplating their now-stony forms.

Up in the balcony, past the room that housed the infamous Wilberforce-Huxley debates, past the collections of stuffed swifts and owls, I found a case filled with tiny blown glass models of anemones. They were not live—anemones are too fragile for that sort of treatment—but they were lovingly spun by a family of glassmakers in the nineteenth century, people who saw their beauty as a source of wonder. They wanted to make these beautiful shapes in part to plant awe in my imagination a hundred years later.

I visit the cathedral of science and the church of Darwin to find hope. There, I connect with the creations of people who believed in the world's possibilities, who believed that life on earth offers the chance to relate. I find faith in the process, faith in the future, and faith in the spirit of survival.

Notes

1. Richard Van Gelder, unpublished letter to George Stofysky, August 16, 1976.
2. Henry David Thoreau, *Walden or Life in the Woods* (Boston: Shambhala, 1992), 8.
3. F. David Peat, *Blackfoot Physics: A Journey into the Native American Universe* (London: Fourth Estate, 1994), 64.
4. Terry Tempest Williams, *Pieces of White Shell: A Journey to Navajoland* (New York: Scribner, 1984), 44.
5. Leonard Shlain, *Art and Physics: Parallel Visions in Space, Time and Light* (New York: Quill William Morrow, 1991), 164.

Cherie Sampson, *Fresh Water-Fire Altar,* 1992 (detail)

From *Broken Island*

Eva Salzman

Lange Eyland, Nieuw Nederlandt (New Netherland), Flora, The Ridings (East and West) Nassau Island, Nahican, Sewanhacky, Sewanhacking, Suanhacky, Sweanhaka (Place of Shells), Mattowak, Mattawake, Matouwax (Land of the Periwinkle or Country of the Ear-Shell), Paumanok, Pommanocc (Land of Tribute), West Egg, East Egg, Ponquattuck, Broken Island

* * *

". . . so white and silvery, calm and pleasant . . . with its long-rolling waves in summer, sounding musically soft against the hard sand; yet how many ship has met her death-wreck, driven on those sands, in the storms of winter."

"Still, there will come a time . . . when nothing will be of more interest than authentic reminiscences of the past."

Please enter the Principality of Ponquattuck, the Land of Water.

Please enter the Chapter of Maquiogue in the Principality of Ponquattuck, to meet some old friends, and reacquaint yourself with this place and its many versions of water.

Please cross the Hemplestead Plains, make your way through the Chapters of Laketown, Bayville, Atlanticville, Waterville, Pacific Beach, Oceanford, Bayshores, Seatown, and pass through the Dictionary of Indian names until you reach Rivertown, the County seat, and finally the tiny blink-and-you'll-miss-it Chapter of

Maquiogue which is located pretty far down island, which is at the heart of it, in its small-town way.

It is preferable to approach by sea if possible, even-though this is hardly the favored means of transport these days.

At every turn, in the great principality of Ponquattuck, you meet an inlet or harbor. Every street in every Chapter—as the towns are called—ends in some version of water:

Bays: with pebbled, seaweed-encrusted shores, its sand sprouting patches of brick-rust iron, pock-marked with crab and steamer breath-holes.

Kettlehole Ponds: one of the many mementos from the last glacier's thump and melt, fed by freshwater springs. And bottomless, according to legend.

Rivers and creeks.

And, of course: The Sea.

That's the big one.

Once visited you take the smell of the sea with you everywhere, for the rest of your life, take the smell of its adornments, the necklace of sculpted driftwood and seaweed pods, the husks of crabs and shellfish and eels.

For the rest of your life, your feet can feel that white sand, fine as talc, scorching at midday and cool at night. Your heels and the balls of your feet are toughened from walking over stones, razor-calms, and burning sand, so that the rest of your life you can walk over broken glass—as you do, as you will.

Even from miles inland you can hear the surf's constant murmur, especially at night when the wind and darkness carries every tidal nuance through the air. The crash, thump, thunder and wallop of more storm-wracked days.

During the blazing high summer days, when the pattern and size of the waves seem more welcoming, warmer, human-scale, and the sound of the breakers softer, more playful, you breath in with the suck of the undertow.

The sea's voice is persistent but changeable: sometimes it is argumentative and barracking, sometimes deceptively soothing, sometimes instructional, and at other times resigned.

Even under the huge but fast-diminishing stretch of Pine Barrens which extends down the middle of Broken Island like a spine of low brush, scrub, twisted and stunted oak forest, even under this dry tinderbox made by nature to ignite regularly, there is water.

Under and over everything: water, water, water.

It all started with a huge glacier inching forward then backward, leaving an outwash, sand and moraine, which is all Broken Island is.

You could point to the sea, how it greedily devours objects and lives, but in fact it may be the water tables flooding Ponquattuck first of all, before the ocean gets to it.

One way or another, it is heading to the bottom of the water.

If you have just come from the Great City, after acquiring a taste for bridges and water, a taste for Broken Islands, you should know that the people around here won't welcome you.

It may be that you cannot swim, but neither can many sailors. It may be that you are afraid of water, but these same sailors or fishermen might call that a healthy respect for the habitat they've never pretended to rule.

It is not your silly and romanticized concept of country folk that makes you unwelcome, but your silly notion that the people of Ponquattuck are a service industry for city people like you. You probably think that Ponquattuck residents exist solely to supply the city's finest restaurants.

That they are there to dig for your steamer littleneck clams, pot your lobsters, basket-scoop your cherrystone clams, pick your samphire grass, tug your clumps of mussels from the knotted marsh grass, and pole-net your soft-shell blueclaw crabs, net your shrimp and scallops, catch your eels, fish your bay snappers and their parent bluefish, fish your sturgeon, your tilefish, your flounder, your striped sea bass. That they are there to pick your watercress, your butter and sugar corn, your beachplums and blueberries and boysenberries from the low brush forests and dunes, your blackberries from the bushes along the road.

To make it worse, you may also feel that, here in the spectacularly beautiful Principality of Ponquattuck, in the Chapter of Maquiogue, things are not as they used to be. They are not the way you remember them, the way you would like to remember them to be. It would seem that the box turtle's slow crackle is not a sound you hear much anymore. You can't recall the last time you've seen that ribboning crossfire from an army of fireflies. Or the blowfish plumping up the nets. Or the local oysters. The small delicate bay scallops.

As if in collusion with the Great City's ironing out of green, the Pine Barrens—the Desert of Arabia as it is called—has been cut into over and over again on Rivertown Highway: car dealers, fast food, mini-malls of pizza places and video stores and 7-Elevens.

Maybe memory has played its usual tricks on you. It may also be that nothing is as unbelievably beautiful here as you thought it was. In which case, you'd rather not know you were mistaken.

It wouldn't be surprising if Ponquattuck forests were filled with voices of the dead: your own personal dead, as well as the growing numbers of the extinct, which you're more bound to notice now than you used to.

There are other familiar and ordinary treasures of sounds and sights, some natural, others natural in a different way: the railroad; crossed phone-lines (convenient for keeping up with the neighbors); a heron sailing across the marshes; squawking geese and ducks; the midday siren courtesy of the Fire Department.

This Fire Department, made up mainly of local volunteers, the fathers of people you knew, has its own well, part of Ponquattuck's treasure of water, which brings you back, always brings you back to the main treasure, the big treasure, the treasure of treasures:

The sea.

And with those seas and bays, more treasures: the smell of rusted hooks and wet; salted docks and rotten fish; the smell of wild rose on the roadside mixed with the smell of melting tarmac and the feel of your bare feet sinking into it; the buzz and thump of motorboats; foghorns and drawbridges; lighthouses; the flap of a sailboat; the plish-plosh of fish; the light lapping waves of the bay.

This is a place where husbands sank in boats just yards from shore, within sight of their home and wife and children—or sometimes even with their wife and children. From these shores the whaling industry prospered and declined.

All of this is stuff you didn't really know consciously, but now you are part of the history too, and the thing about history is that it always depends on who's telling it and for once, you'd like to tell your own small version.

Every time you are near the sea, you grapple with an intimate memory which never quite arrives but is always in the process of appearing.

Here's how you return.

You pull up in a perfectly ordinary village, on a perfectly ordinary Main Street. It is late morning, and the town seems empty. You almost expect to see horses and carriages, but this is not very far back in time.

In fact, it might even be slightly in the future. It's hard to tell in a place like Ponquattuck, where the past and future are so entangled.

It may be spring, late spring—out of season anyway, the streets are still empty of city people.

In any case, Maquiogue is not one of the more upmarket Chapters in the Kingdom of Ponquattuck. The local realtors can't charge top prices here in Maquiogue.

In Maquiogue there are no cafés in which to sip Iced Lattes, eat rabbit food or goat cheese. The appeal of Maquiogue is its sleazy and downmarket feel, places like Frankfurter Beach.

For regular guys playing pool and downing a few beers, try the Maquiogue East Bar.

For those more upscale, try the Casa Nuova with its crystal chandeliers, marble bars and wallpaper glitter, with its nonstop poker game in the basement and its middle-aged piano man who wears a toupee and is busting out of his tux. Stroll past the peroxide barflies crammed into short, tight skirts; take home an overweight married man with slicked-back hair and gold medallions lost in a twisted mass of chest hair.

No, Maquiogue is very different from the classier Yeasthampton (as it is christened by Mrs. De Robben) and even from Wastehampton (as it is christened by Mr. De Robben) which feels superior to Maquiogue in the way that cons in prison create their own hierarchy of lower and upper crime.

And then there are the "Hamtowns," which run down the spine of Middle Island, between the North Pork and the South Pork.

As they were, are christened by Mr. and Mrs. De Robben.

Depending on the timing of your visit.

Mr. and Mrs. De Robben run the local stationers—ran the local stationers—and you would definitely like them more than most of the other adults around here. You could imagine her smoking a joint, you could imagine him as your favorite grandfather, puffing his pipe, watching intelligently through the sweet, curling smoke, but never criticizing too much, never telling you off. He would listen carefully to everything you say, and nod in that slow careful way which shows he listened and that it's important.

The jitney stops at the Chapter's edge, and you climb out and walk toward the center of Maquiogue. You walk past the gas station Esso, the Realtor's Office with its electric turquoise hoardings, the tiny Post Office with its brass honeycombs of post boxes, the Fire Station, the Diner, the Deli, Maquiogue East Bar—for the locals—and Silver's Stationers run by the Mr. and Mrs. De Robben (whom you would like).

You pass the two churches, one Canolothian—that wonderfully sanctimonious and censorious sect housed by a newish, and spare red-brick block—and the other Luserian, an older glorified whitewashed clapboard house with a toy-like slatted steeple, out of which wafts electrified rinky-dink tunes, which fill you with a warm glow of nostalgia.

They don't really like each other very much, these two sects, even if they are so geographically intimate, and if each congregation gets an earful about loving your neighbor.

Everyone knows that listening to such sermons is as good as actually putting them into practice.

Nearby is the meeting house for the Canolothian Daughters of the Revolution, which is infinitely more exclusive and less well attended than the meeting house for the plain Daughters of the plain Revolution.

You won't be invited into either place, so keep walking. You're a city girl, however much you feel this to be your town, your Chapter.

Keep walking, past the cedar-shingled houses, some unpainted gray, simple farmhouse boxes, the larger, more well-to-do whitewashed clapboard topped with cupolas and dormer windows peeping out above, flowing with net curtains. Even many of the smaller, meaner houses are encircled by a wide trellised porch.

At night a lantern or electric candle is framed in the windows of the houses, which makes you think of skeletons in closets.

Through one window, behind muslin curtains, a television flickers. A Pontiac sails through the street. A Chevrolet, a Buick or one of those "tin lizzies," those huge and low-slung boats navigating the Main Street of your dreams. Somebody's father, somebody's lover, or a nobody's husband.

It feels as if you are holding your breath, just outside the borders of some painful, yet strangely pleasurable memory. You walk on out to the edge of town, to the 7-Eleven which always used to sell your cigarettes.

"Yer lucky I still had 'em," says the store owner, a very short but stocky balding Polish man with a moustache. "No one buys 'em."

He throws down the packet onto the counter.

The fire station siren goes off. You pause a moment, out of habit, the way everybody does when the siren sounds, to listen to its drawn-out ugly-sounding scale, the low moan sliding up to a high wail, before falling back down to that broken hoarse rumble.

"Well, I buy them," you reply. "You can save them for me."

"I've never seen you before."

"You're jokin'," you say swiping the pack and cramming it into your pocket.

After all those years of working to belong.

Anyway, he's a little confused, because you're acting like you know this place like the back of your hand. During high summer, strangers and their strange cigarettes would be expected. The rest of the time, for any normal person, the poison is menthol Salem and Marlboro.

Now you can hear the more distant midday siren of Rivertown; time varies in Ponquattuck from Chapter to Chapter.

You wait for the siren repeat—the storekeeper's sort of waiting too—but none sounds. This means it's midday. This means no fires, or no fires that anyone's yet discovered.

"Shit, it feels like the end of the day," you say. "It's only noon."

"It don't feel like the end of the day to me. Wish it did."

Outside you light up and find out how right he is about no one ever buying your cigarettes. This one's an antique, judging by the taste of it, the most disgusting piece of crap you've ever tasted.

So get off your high nostalgia horse about the old days.

Get walking again, back into town, spying through windows along the way, trying to catch glimpses of people at dinner or in their television pose. A woman hurries out of the Post Office and turns a corner before you can ask directions. She's familiar, someone's mother, maybe Allen's mother, Mrs. Sayre, the husband's nobody wife!

A teenager hurtles around the corner on his bike and is gone; he's familiar too. Allen Sayre maybe, younger than he should be. But this is even more impossible than any of the other impossible things, even in the kingdom of Ponquattuck—even on Broken Island—there's no saying what or how or with whom.

Not every kid you see belongs to your life. Get real. Except, it's hard to do that in this place.

A little farther an old woman in a wheelchair is leaning forward and murmuring to a toddler seated in a stroller pushed by a rough-looking young man.

If Allen Sayre was too young looking—or, rather, the kid who looked like Allen Sayre—then Billy Tuttle looks older than he should be, much older and, boy, has he changed. His angelic look is gone, his blonde hair now long and greasy, his Kurt Cobain T-shirt sleeve rolled up around a pack of Marlboros. He has a skull-and-bones tattoo and there's a cigarette dangling out of his mouth—a Marlboro of course—the smoke curling up into his squinting baby-blue eyes. They are almost scary, those sky-blue eyes.

Well, well, he made it. With a child to carry on those genes, a child to mold and shape. A child to take after him the way he took after his father.

He doesn't see you; not that anybody would recognize you; everyone's the wrong age, including you.

As you walk by, you catch the woman's voice:

"Heya, little Frankie baby," she coos, her wrinkled hand under the toddler's chin. "This is your Aunt Jane. You remember your granddad's sister? Two years old and you ain't never met your own Auntie before."

This last, said in a voice like a sharpened blade, is directed up at Billy, who couldn't give a flying fuck. He's waiting for her to be through, but she's got his nasty genes; you always expect it to have gone in the old, to have softened into something good and wise.

"You been inside again?" she barks at Billy who's looking across the road.

She snatches the butt dangling from her own lips, hurls it into the street, where it hits the sidewalk with a small cloud of sparks, then she wheels herself off vigorously.

"Fuck's sake," you hear Billy mutter under his breath.

A thin-looking girl comes out from the store across the street, hardly looks both ways before she saunters across.

"Take yer time," says Billy.

"Okay," she says coolly, puts the buggy in motion with a jerk, causing the child to yelp.

A warming family scene.

You buy a newspaper at Mr. and Mrs. De Robben's stationary store, that landmark: Silver's Stationers.

She, who used to be so lively and easy for kids to talk to, who burned incense and told fortunes, fumbles around for thick glasses and even then can't find some bit of paper she deems necessary, for some odd reason, to sell you a newspaper. It's a relief she doesn't recognize you.

Mr. De Robben is sitting there fingering the yellowed pages of an old *Maquiogue Herald,* in which appears an article of his, an article about boats or bridges, an article about nature and the environment, about global forces, about pesticides and the bad guys: about Broken Island.

You walk past the Maquiogue East pub, and a bleached-blonde woman you almost recognize is perched on a barstool, the bared skin of her midriff soft and beginning to fold.

You turn off by the Fire Station and walk down Bay Avenue turning left at Foster Crossing and then right again on the dirt road, which takes you past the Kovskys. The place has been painted and the lawn is immaculate. The screen porch looks new: no jagged holes or tears at the edges or dents from a boy's consistently misthrown ball. There's a shiny new station wagon in the driveway.

The swing is gone. There is no ice cream truck.

Not yet ready to head on, you backtrack to Bay Avenue, turn left and go as far as you can, which isn't far. The road peters out into a dock on the expanse of still water which is Shinnewockey Bay. An osprey sails over a spit of land opposite and settles at the top of a dead tree jutting up from the end. An imperious swan slowly makes its way past you.

In the water the minnows flip and flutter, and the water laps against the stilts of the dock.

It's not as if there is anything spectacular to see—no majesty of terrain or splendor. It's just a mirror for the sky, the gentle push and pull of its slow tide, the equally slow swoop of egrets and kingfishers, a blue and wide peacefulness.

It's a warm dusk and you can hear the purr of motor boats though none are in sight. No one cannonballs off the dock. It's not warm enough for the hordes of summer swimmers even if you can feel the absent August baking heat. No one screams bloody murder as they're dragged across the hot and broken tarmac to be hurled into the water like a sack of local potatoes. 'Round the tip of the near land, a sailboat glides into view, silent as a cat.

Aside from all this, it's an ordinary place.

So you turn back and walk through overgrown trees and bushes, until you arrive at an old white farmhouse. Its nearby garage is twined with a mass of trumpet vine, not yet in flower.

At a slightly farther distance there's a small shed-like building that appears to be bursting with junk, and dotted around stand portions of gray cedar fence, over five feet tall, also congested with weeds and branches, probably a dog rose or climber of some sort. Buckling bits of the broken fence jut up from the tangled undergrowth, like marooned remnants of some grand, crazy plan which never had a hope of completion.

The paint on the weathered shingles is peeling and crumbling, eaten away by salt winds and hurricanes and baked by the heat. The whole structure slumps and sags. Surely nobody lives in this place. Rosettas of emerald moss and mold flourish 'round the windowsills, cascade down the sides.

You move toward the house, crouch down, and carefully peer over the windowsill, looking at the family gathered there at the table, your family. There is your sister, your father and mother, a large raggedy-looking goofy cartoon of a dog, with its rough and shaggy coat, its tongue lolling out sideways.

Someone is yelling at someone and the dog is helping himself to the food on the table.

When you go in, nobody is surprised to see you; nor do they ask where you've been.

All this distance to have come no distance at all.

"Oh, hello everyone," you say with some sarcasm since, as usual, they don't appear to have missed you.

"Please," your mother sighs, "not now."

They are all very busy. Anyway, it's not like you'd written.

It's okay. You won't be staying long. You didn't go through all that time for nothing. You pass through the door on your left, open it, and find yourself back at the first door again, entering the kitchen.

"Oh, hello everyone," you say, pretending everything is normal.

This time you retreat and try another door, but it doesn't work. You just end up in the kitchen, approaching from another angle.

The third time you try it, you end up outdoors, walking down the path to the pond, and then to Shinnewockey Bay. About ten feet in front of you, sticking up from the surface of the water, there's a door. You wade out fully dressed, open the door, and step through: back into the kitchen.

Whatever you do, you keep returning to that same room, to those same people.

Were you to go to the other side of the universe, you'd still end up at this same place, at that kitchen table, with the same pictures hanging lopsided on the walls, the same arrangement of mildewed furniture, the same dog, the same answers.

Here is your poisoned chalice. To preserve it all—the birds, the creatures, the grass, the sea—is to undergo what must be a constant trial. Don't you *want* to be held frozen inside of it, like dried flowers under glass, like the little snow scene in a glass globe?!

The creatures, the grass, the sea.

The sea.

There is nothing so ordinary as the extraordinary Principality of Ponquattuck.

And it is often the case that the inhabitants of a magical Principality do not recognize its magical qualities.

They may be blind to its beauty so as to be able to live there.

Peter Goin, *Coyote Howls at Old Divide Trading Post, Colorado* (detail)

January

David Petersen

> *My prejudice is pretty general, far too broad and sweeping for any racial limitations.*
> —Hunter S. Thompson, *Fear and Loathing in America*

My year-round home is a little cabin on a big mountain in Colorado. By January here each year, after weeks of foreshortened days and long, drowsy nights, semi-hibernation becomes a comfortable routine. After this fashion, we—my wife, Caroline; our dog children, Otis and Angel; and I—emulate our shaggy neighbors in the surrounding woods, snoring contentedly in their snow-covered dens. While we can't and don't care to go that far, C and I do practice semihibernation in winter, retiring much earlier at night. On winter mornings, we tend to rise lazy and late and only then at the insistence of Otis, who rouses as the first beige of dawn seeps through the bedroom curtains, asking to be let out.

After dragging myself away from Caroline's magnetic warmth, I pull on a robe and moccasins, scratch Otis's ears, and let him out to conduct his morning piss patrol around the boundaries of our pauper's estate . . . take a leak myself (outside in summer, inside in winter) . . . stir the ashes in the woodstove, top the glowing embers with a few kindling sticks of fast-lighting aspen, add some larger wood, open the damper and close the stove door . . . put on a pot of coffee and dress while it perks.

After washing hands and face in fifty-four-degree well water (we have hot water, but I need the wake-up shock), I tune the radio to NPR's *Morning Edition*, which occasionally surprises me by turning out to be *Weekend Edition*. As a self-employed work-at-homer, one day is much like another, and I occasionally lose track. For good and bad, weekends mean nothing here. Weekends and holidays, in fact, are my most productive workdays, when the phone rarely rings.

When Caroline and I first moved up here to the mountain—on April Fool's Day 1981—to live out our *Mother Earth News* back-to-the-land dreams, we had no telephone. And for the first few months, no electricity. Boldly challenging my lifelong fear of debt, we had signed a $5,000 owner-carried loan (over three years, plus $500 cash down) to buy this wooded bit of mountainside at a far back corner of the oldest summer home subdivision in the county (platted in 1966), which, in '81, was still mostly empty. So we took the leap and bought the land and after scraping together another $400 cash to buy a 1950s vintage 8 × 24 travel trailer, we set it up, cleaned it out, and moved up here from town. We became the fourth year-round household on the mountain, with none of our neighbors within sight or hearing. No penned-up or chained-out dogs barking incessantly. And best of all, no delinquents running wild on ATVs. Quiet, private, semiremote. A poor person's paradise. That was twenty-one years ago, fifteen years before the inevitable Californicator invasion began and compared to now, a heavenly interval indeed, if never quite easy.

Consequently and sadly, this place then and this place now bear little resemblance. As "progress" overtook the region and real estate values soared, more and more people moved in around us. Summer homes were upgraded to year-round residences. New homes were built. Now dogs bark incessantly, and kids on motorized vehicles are a summertime blight. Adjoining larger parcels of private land have been logged and overlogged, and the wildlife, myself included, feels the painful pinch. "The thing you have to watch out for with progress," advised my dear friend A. B. "Bud" Guthrie, Jr. (who won the 1951 Pulitzer Prize for fiction), "is that once you have it, there's no going back."

So true, Bud, and so very sad. Yet this is home and we cannot leave.

Back in '80 and early '81, while still living in town, we called our final rental house the Slave Quarters due to its cramped size, tarpaper exterior, and placement behind one of the biggest nineteenth-century Victorian mansions on upscale Third Avenue. Yet even the Slave Quarters was an improvement over our original Durango digs—a second-floor room with kitchen in a cheap motel in a wind-scoured canyon a mile west of town, where the winter sun never reached and the parking lot ice never melted. After a couple of months, we moved from there to the Slave Quarters. Then a few months later, we moved again. To the mountain. To here. And here we got along just fine—without electricity or phone, that is. If it weren't necessary for a 240-volt submersible pump to lift water from a 278-foot-deep well, we might still be juiceless today. No indoor plumbing back then either and no biggie. After all, for millions of years and until just recently, people everywhere got along without flush toilets, as millions worldwide still do today, though precious few by choice.

Of course, when you have the option at your fingertips—"*could* go without electricity" versus "go without"—it's no contest, and Caroline and I do love music. We also enjoy a synchrony in musical tastes, far-ranging but particular (the ten-cent term is *eclectic*) . . . a happy discovery made one steamy night in the summer of '78, in the big white house at 101 High Drive in Laguna Beach, California, where I roomed with two fellow former marine chopper pilots who, conveniently, were away for the weekend. It was there, after dating politely for way too long, that Caroline and I finally got together, to the stirring accompaniment of Firefall's *Livin' ain't livin' no it ain't livin' alone* replaying endlessly while we—young, uninhibited, athletically healthy, and deeply in lust—were otherwise occupied, confirming and reconfirming the truth of Thoreau's "awful ferity with which . . . lovers meet."

By daylight, the wear was so severe that I had to replace both the album and the phonograph needle. But the union, between C and me, remains fresh and young.

"What kind of music do you like?" someone occasionally asks.

"*Good* music," C and I answer as one.

And good music—delivered via radio, cassette, CD, and our prized collection of 123 vintage vinyls (including the replacement *Firefall*)—is generally enough virtual entertainment for me. Aside from the occasional bout of motel-room channel surfing, I haven't watched TV for more than thirty years and don't feel I've missed a thing. Sprawled on my deathbed—which, if I have my way, will be a carpet of moss hidden deep in some secret aspen grove—reflecting back on my finished life, I doubt I'll exclaim, "Damn, I wish I'd watched *more TV!*"

For much the same reasons that I don't watch the tube, and Caroline rarely does, we avoid commercial radio. Nor do we often read newspapers, national or local, or news magazines. Consequently, aside from word of mouth passed among friends, our primary tap into what's going on out there in the "real" world is public radio. NPR's evening *All Things Considered* and *Morning* and *Weekend Editions* are our favorites, and best of all is the wise and wizened voice of preeminent political commentator Daniel Schoor, who is to radio as Hunter Thompson is to print, both of them providing some of America's most honest and insightful political criticism. *Democracy Now, This Week in the Media* and a few other sharp-edged "cultural conscience" productions, from Pacifica and other public radio venues, also are winners here. But even public radio news (Dan Schoor excepted) quickly becomes repetitive, boring, and largely pointless when "big news" (meaning always *bad* news) is going down. Consequently, a few minutes of radio news once or twice a day—I try for the top-of-the-hour quickie headline recap—is news enough for us.

Of far more interest here is the chittering, gossipy news being broadcast from just outside the cabin by station WBF (winter bird feeder). Regular commentators there include stately Steller's jays in powder-blue vests, sooty-headed, high-crested, bright-eyed (an illusion owing to their white "eyebrow" streaks) birds as raucous and temperamental as they are lovely; black-hooded Oregon (winter only) and gray-headed (year-round) juncos; nuthatches (white-breasted, red-breasted, and pygmy, tree-trunk-crawling woodpecker wannabes every one); plus tiny piney siskins and cheery chickadees (mountain and black-capped). And all of them crowded together on the snow below the recycled tonic bottle seed dispenser suspended from a slender aspen.

Farther away and more aloof, casual formations of common ravens cluck and caw bemusedly as they seesaw through the icy sky, headed who-knows-where for another day invested wisely in eating, playing, and thoughtful conversation. From the creek valley below come the piercing *Maag-maag!* pronouncements of black-billed magpies, visually striking with their piebald mix of iridescent blue-green wings and long-forked tails, snow-white shoulders and bellies, inky breasts and hoods, every inch of them bright and gleaming. This so-called weed species is increasingly abundant here in winter, though we rarely see 'pies this high in summer, unless there's carrion around.

And so it goes, early every morning, after checking in with the birds, that I shuffle over to Angel-dog's bed in the quietest corner of the living room, and sing her some silly wake-up ditty—whatever oozes from my groggy brain. Happily, Angel is musically uncritical and grins and pants and stretches in appreciation.

With Angel roused, I open every curtain in the cabin, of which there are many, thanks to Caroline's construction instructions back when I was cobbling this shelter together, a mandate that cost me a lot of extra work at the time but since has proven its worth, as the light and the views provided by all those windows help alleviate what otherwise, for C at least, would be a claustrophobic darkness in this woody cabin cave.

And finally, every winter morning, when the coffee is good and black, I pour two mugs full and take one to the Queen in Waiting, reposing prone in her quilted throne, covers over curly locks, another angel sweetly feigning sleep. After kissing my wife, I let Otis back in, let Angel out, let Angel back in (she's old and fast), feed both mutts, return to the living room, plop down in my chair to drink coffee and read for a few lazy minutes, with radio and/or birds softly chirping.

And so it goes on a typical winter's morn, here at the postmodern, technoprimitive Petersen digs.

Several among our big-city friends think and speak of Caroline and me as wilderness ascetics. And as viewed from an urban high-rise, I suppose I can see their point. But ascetic self-sacrifice has never been our thing. We seek merely to control our own days and destinies by controlling our material desires; to straddle the edge—geographically, culturally, and spiritually—between the human and natural worlds, selectively gleaning the best (for us) from both. And we've pretty much made it work. We can walk out our door and up the mountain and enjoy de facto wilderness, pulsing with wildness and life, almost instantly. Or, in one strenuous day's hike from here, fifteen miles, we are in the 488,200-acre Weminuche Wilderness, the largest and (my opinion) loveliest publicly owned commons in the American Southwest. Those same fifteen miles in the opposite direction put us on Main Street of clang-bang Durangotown, where your money is always welcome. We go one way, wilderness, and another, resort-town cacophony, as we choose. Edge dwellers.

Same with technology. We take full advantage of what works for us—good music pumping from a good cheap stereo, a low-end computer to earn our daily sourdough, a cheap VCR and an embarrassingly big screen to watch it on (the latter a gift from our good friend Hippie George, kept hidden beneath a trout-print blanket when not in use, which is most of the time), electric lights, appliances and well pump, small-capacity water heater, telephone (no stinking cell though, never ever), a brawny chainsaw for getting the firewood in, our only source of winter heat—and so on, everything we really need. And to hell with the distracting, slave-making rest.

We like it fine this way.

No screaming alarm clock to start each day. No rush to shower and dress; daily uniform for both of us is T-shirts, sweatpants, moccasins (winter) or sandals (summer), and, for baldheaded me, a baseball cap. No force-fed fast food so-called breakfast. But like most other folks these strung-out days, I must commute to work—about twenty feet, from the cabin to the eight-by-twelve storage shed/office that Caroline calls the Outhouse ("appropriate," she reasons, "to what's produced within"). Aside from C and the dogs, I'm my own boss, and unless I'm on deadline, when there is little else *but* work, work can always wait, a little while at least. Especially on frosty winter morns.

Second cups of coffee poured. Caroline up and bustling about. And as sometimes happens, I find myself grudgingly captivated by news I really don't need to hear, in this instance an interview with some mad Harvard scientist bent on human cloning—not merely stem cells and organs, but the whole-body shebang. *Just* what we need. More hubristic God playing. More people. More overcrowding and self-poisoning and unnaturalness. More quantity and less quality in every way, increasingly every day. More of less, all of which we celebrate, uncritically, as sacrosanct progress.

According to the Worldwatch Institute, some 20 million people die from malnutrition each year, while another 1.2 *billion* don't have enough to eat. Yet some among us want even more—particularly if the additions can be perfect genetic copies of our own perfect selves. (Was this also God's vanity, in creating Man after His own image?) Nor, in my peculiar opinion, should we even be cloning organs to prolong human lives, as if death were all that bad. I mean, if heaven is eternal paradise, why work so hard, spend so much, to try and delay the promotion? If I were a True Believer, I'd be up there already, right alongside the Reverend Jim Jones and his fallen flock.

Before I grow angry—which solves nothing—I turn off the radio and restore my equanimity by petting the dogs and kissing the Queen (again) before ducking out the door for another ten-second commute, another leisurely day of what passes around here for work. No complaints. No regular paydays or perks either. That's the trade-off.

Winter: so lovely and so quiet; the most restful of seasons (between blizzards), custom-made for loving, sleeping, working, reading, walking (always, there's the walking), laughter- and music-filled evenings with friends and thinking, always there's the thinking.

And so begins—and before we know it, ends—another January day.

And another.

We wake this morning to fresh snow on the ground, relieving, if not ending, a weeks-long drought. Around the cabin, these few new inches of white politely hide the unsightly yellow snow (my own as well as the dogs'). Out in the woods surrounding the cabin, the thatch-work boughs of hardy evergreens—white fir, ponderosa pine, Douglas fir, blue spruce—hold tons of suspended snow. Sometime during the night, while we all slept, the silent storm came and went, moving northeast, bound for the alpine peaks of the Continental Divide, thirty miles from here. Typically, if we get inches here, the Divide gets feet. And when we get feet—well, at least the skiers and local businesses are happy. Meanwhile, up shoveling snow off the roof at midnight (to keep the cabin from caving in), I am fantasizing an escape to Tahiti and pondering the hereafter (Tahiti would suffice) should my ticker run out of tape.

As usual after a nocturnal snowfall, the morning is bright, quiet, clean, and cold, the forest glistening and pure. This is Martin Luther King Day, as I am repeatedly reminded by the fatherly voice of *Morning Edition* host Bob Edwards, speaking to me live via the personally incomprehensible techno-magic of frequency modulation, all the way from our nation's capital.

As the sun climbs the pristine sky and slowly warms the air, the snow from the night's gentle fall—wet, heavy, and frozen—begins to thaw and crack and, every now and then, comes crashing to the ground in a tree branch avalanche: *Whump-whump, WHUMP!* A stroll in the woods on such a morn as this is guaranteed to be a bracing experience: button your collar, hold onto your hat, and brace yourself to be clobbered, engulfed, even knocked to your knees by any number of utterly unpredictable tree branch avalanches. Happily for me, and by her own decree, Caroline is the morning dog walker here, no matter the weather.

And so she goes and comes. This morning, in order to break trail in the new snow, she wore snowshoes (not a glamorous way to travel, no matter what the slick ads portray). Over midmorning coffee (I've worked three hours by now and according to my labor contract with myself am entitled to a coffee break), my partner in this life of small adventures recounts the high and low points of her walk. While following Otis (Angel's hard-walking days, sadly for us all, are over) along a twisting game trail—visible to the attentive eye as a shallow linear trace in the snow—Caroline noticed a pair of tiny, lizardy feet projecting from a lump of white beneath a Douglas fir. Grasping the bony digits, she gave a tug and out popped—a chickadee, not yet stiff or even cold, but convincingly dead. The sparrow-sized bird apparently had been avalanched from its perch and buried alive, just minutes before Caroline's arrival. Imagine: You are this little bird, patiently enduring the long, frozen night huddled among the sheltering boughs of a big friendly fir, your downy underfeathers erected for insulating warmth, waiting for the snowfall to end and the morning to dawn. Finally, here comes the sun, caressing your pewter breast and warming your pea-sized heart. Any moment now, you'll wing away for another day of freedom and adventure.

And then . . . *Whump-whump, WHUMP!*

You are trapped, inverted; you cannot move or breathe. All is dark and cold again.

Nature: It isn't always pretty, but it darn sure always works. And rarely is there waste. In this case, Caroline left the little feathered body lying on the game trail, where some hungry scavenger—raven, magpie, coyote, fox—soon will happen across it, perhaps already has, recycling death into life.

"Born again," that little bird, as close as it likely gets for any of us.

This morning's telling of Caroline and Otis's excellent adventure reminds us both of a similar, even more chilling drama, which C not only witnessed but was also part of, back in late November, during a sudden blizzard that caught her out and unprepared. Accompanied by Otis, my wife had gotten only half a mile above the cabin; outbound on a midmorning walk, when a ferocious squall of blowing snow came blasting in without warning. Even here in the comfort of the sturdy Outhouse, it was an impressive storm. Visibility outside closed down to nothing, day into night, and the wind blew so viciously I became concerned that the fork-topped ponderosa on the hillside above me might snap like a wishbone and ruin my days forever.

They're not called widow-makers for nothing.

But mostly I worried about Caroline, out there alone with Otis.

But nothing to be done 'til it's over. In such a blow there are no tracks to follow. A shout for help becomes a whisper. And C, spontaneous and independent, never tells me where she's going, assuming even she knows. And so it was, this bad blizzard day, that my wife was on her own to find what shelter she could—predictably, the leeward side of a thick-trunked tree—and hunker down with Otis until the squall had passed . . . or, should it storm all night, until the spring thaw, when I'd find their frozen corpses emerging from the melting snow, like a pair of chickadees.

"From the instant the storm hit," she told me on her return that spooky November day—still zapped on adrenaline, talking fast, gesturing wildly—"I was blinded, could barely see Otis just a step ahead. I yelled for him to take us home, but he was blinded too. The wind blew harder, and the snow stung my face so I ducked beneath a Doug fir and Otis had just crawled in beside me when I saw a yellow-orange aspen leaf come hurtling down from a limb right above and go skittering over the snow, then flutter back up a ways only to be knocked down again, two or three times in an instant, right in front of us. By the time I realized it wasn't a leaf at all but a bird, a junco I think, it had disappeared, buried in the blowing snow. I felt around but couldn't find it. It was heartbreaking, and there was nothing I could do."

Heartbreaking, you bet. Yet compared to our human-made world, all the daily "tragedies" and "cruelties" of nature seem downright benevolent. Out there, at least, there's a plan.

With the midmorning coffee break and postwalk tell all done, and a token bit of work accomplished, I make ready for my twice-weekly, thirty-mile round trip to town. (For the record, I work way more than the national average forty hours a week, and without overtime pay. But—*Hallelujah!*—working for myself, I can *choose* my hours and days, so that work fits comfortably *within* my life rather than ruling it.) Today's town chores, the usual, include a load of laundry, market, post office, and fleeting conversation with townie friends: Fred at the King Center Laundry; Linda (a fellow mutt-nut) at the Albertson's check-out; good old Hippie George, dispensing used books, collectible junk, fine antiques, and cynical good humor from his cluttered Southwest Book Trader, a block down Fifth Street from the tourist-trapping Durango & Silverton Narrow Gauge Train depot. There are other town friends—photographer Branson Reynolds, Doc Dave Sigurslid (our personal physician and maker of artful wood hunting bows), Ken and Ann and Danielle, barrister Steve, and Professor Grigg . . . far too much society for any one day. Caroline, being even more reclusive than me, can be dragged to town only, and not even always, with the bribe of a dinner out. And I really don't mind so much—doing the town chores, that is. A couple of hours adrift in clamorous civilization, a couple of times a week, make me eager, and thankful, to get back home.

Martin Luther King Day. It's not the Right Reverend Doctor's actual birthday, of course; that's January 15. Rather, today is the floating three-day weekend virtual version. As per tradition, off and on all day, the local NPR radio station, KSUT (owned and operated by the Southern Ute Indian Tribe), airs programming appropriate to the occasion. Soulful black gospel music, mostly, including a particularly stirring choral a cappella of "We Shall Overcome."

As I motor down the mountain, townward, in my trusty '79 Toyota 4 × 4 minitruck—123,000+ miles when I bought it, several years ago, for $3,200 cash, and now, with a $2,500 engine rebuild, well past a third of a million miles and still humming sweetly—motoring along, watching as always for wildlife on or near the road and listening to this poignant, moody music interspersed with moving tributes to King delivered by fellow race-reform warriors of the tumultuous '60s era, I grow increasingly maudlin, a pushover for Laura "Octoroon" Love's haunting a cappella-segue-to-rock delivery of "Amazing Grace," followed immediately (the old one-two) by Iris Dement's tragically true "Wasteland of the Free."

What was it my old friend and mentor Edward Abbey had to say along this line? Something poetically powerful as usual—something about "those who do the world's work and are never paid enough and never will be, and they rise and are beaten down, and rise again and are beaten down again, and always lose."

And *always* lose, even as the rich always win. At least, such has been the way since the advent, some ten thousand years ago, of agriculture and surplus production and what novelist-philosopher Daniel Quinn (*The Story of B*) terms "locking up the food" and the unholy hierarchy of Big Men and hired guns and wage enslavement we've come to know as civilization.

I'm no bleeding-heart liberal, mind you. I go out in the woods and kill things, occasionally with my bare hands, rip their steaming innards out and eat their bloody flesh, cooked or sometimes raw. Exactly as humans are meant to do. Just last September, indulging a long-held Neandertal urge, I devoured a hunk of still-warm elk flesh, freshly carved from the backstrap of an animal I'd just killed with bow and arrow; it was mild and delicious and I suffered no ills. Nope, no bleeding hearts here. I even endorse capital punishment for inhumanly hurtful, utterly worthless, and immutably guilty villains. Yet in spite of my calloused, jaded, unforgiving nature, hearing this moving music in honor of America's ongoing internal struggle for equality and basic human rights—which struggle Martin Luther King embodied—I am overwhelmed with empathy for those who do the world's work and are never paid enough and never will be, and rise and are beaten down, and rise again and are beaten down again, and *always* lose. As one who has it easy, I am consumed by shame and sorrow. And so little hope.

A bittersweet bile of emotions floods up from my guts, pools in my eyes, and streaks down my weathered face, clouding my vision, making it hard to drive.

I think again of my saintly mother, at peace in her early grave, and of my father, alive yet forever beyond my reach. I think of gentle Angel dog, tired and waiting, so gracefully, for the hole I've already dug. And I think of the buried chickadee, of the storm-tossed junco, of the bright shining lie called security.

Accompanied by these helter-skelter musings, I power my way through town, chores, and traffic in a hectic two hours. (Anyone who sits in a coin laundry and watches their dirties go round, or stares at some inane shit-come on the ubiquitous laundromat big-screen, when there are infinitely more worthwhile things to be done with that time—other chores, the library, a relaxing side-street walk—that one either lives in a neighborhood where unattended laundry is at risk of disappearing or is hardly the sharpest pencil in the can.)

Trucking homeward, uphill all the way and halfway there, the woes of the world still screaming like chainsaws in my head and without a heartbeat's warning, a huge dark shadow comes swooping down from my left . . . disappears below my vehicle's hood and close in front, as if it will pass beneath . . . flares back up, in full frontal view, its belly and open yellow talons facing me, spread-eagle, two feet from the windshield, three feet from my face. This imposingly beautiful beast, its wings spanning seven feet, shrouds my forward vision, blocking out the sun . . . hangs there for an instant, suspended in time and space, branding itself in memory, so close I can see *into* the twin black teardrop nostrils atop its gaping yellow beak. And flaming in the bismuth pupils of its amber eyes is . . . what?

The grand and lovely mystery of life and death on Earth.

Suddenly I *know* how a rabbit feels in its last instant of consciousness, with winged death descending.

Jesus H. Christ!

Even as I'm registering these machine-gun images, I stomp the brake and swerve hard right—useless, tardy reactions. But the agile raptor is faster and just fast enough to save us both, tilting kited wings and lofting gracefully up . . . as I slide unscathed beneath.

Committed irrevocably to a semicontrolled skid, I screech off the side of the road, stop at last on the shoulder, pop the stick-shift into neutral, yank the hand-brake, and leap out—just in time to see my angel of near-death flare wings, extend talons, and settle to perch at the top of a cottonwood, like a black and white Christmas ornament. Aloof, surreal, magnificent.

A bald by-god *eagle!* Our national bloody *emblem!* Right in my pretty *face!* On Martin Luther King Day!

I contemplate with gratitude the slowing, uphill grade and patchy black-ice threat of the narrow winding road, conspiring to slow my speed. What if I'd been going faster—the limit through here is forty-five and I often drive fifty—and had struck and killed the eagle? How would I deal with *that* wee irony? Assuming, of course, I survived the crash sure to follow.

Looking at that proud beast now, glaring from its leafless perch, across the silvery sliver of frozen Florida River, our bug-eyed encounter of moments ago seems already forgotten. Already, the feathered neck is on the swivel, the searching eyes and hungry mind back where they belong, on the hunt.

Suddenly, all the Sisyphean weight of the human-made world and this poignant day lifts from my chest and floats lightly away.

Nature, for those so attuned, provides a reliable antidote to the despair and insanity of human culture. When the last of healing nature is gone, murdered by greed and gone forever . . . what will become of us then? Life without nature—without wildlife, wild places and wildly natural people—would be, for me, too bleak to bear.

Are there eagles in eternal paradise? Are the heavenly fields of Elysium peopled with dogs and trout and shambling bears, the keening of hawks, aspen leaves rattling in a sweet autumn breeze? Will young couples laugh and make love amid spring wildflowers, and oldsters, holding weathered hands, remember and smile or weep?

If not, then to hell with heaven for me.

I don't know where we're going, but I do know where we've been. And where we are right now.

Thank you, friend eagle.

Rest in peace, Dr. King.

Aaron Holliday, *Untitled* (Baby Fish), 1995. Courtesy of Ricco Maresca Gallery

Dave Brubeck's Garden

Ricardo Pau-Llosa

This is the Florida I might have seen before
eyes arrived. A vision: nature, a scattering
of barbed leaves routed, a trunk fallen,
lichened, mossed, drenched, bromeliad laden.
The rich tangle for its unwitnessed self
lived, or did it? Does the stark twittering
of beaks possess its depth though no one delve
into its how's or when's? A musical score
dozes on the piano, awaits its spring
and with it the course of mind that equals action.
Through bloom-flood fertile mores
move, roost, chant: witness meaning.
The dream of nature too's a human thing;
mind is nature's destiny, its self.

Progress

Ricardo Pau-Llosa

They are the flood's fingerprints—the lizard
scurrying across a wall, the fly's buzz gelled
in hover, the spider's hairline drift,
the vulture and the albatross and all
the kited wings, all the living things
that bring the sea's unleaden habits
to air and land. Their floated
masteries would a normal practice be
had the rains endured and the whirling
arks never keeled. It is no accident
that when men fled the dense quotidian
to rub the empty stars, they wombed
their weight in freedoms
of untroubled space. Suit bubbled,
tubed to air and duty, they bowed
to the mosquito lighting on a pond,
the mantis and the jay. They homaged
the bat's pendulum and the ant's intact dive
from the canopies where serpent and monkey
mirror each other's coil and hold.
The flood, maligned in murderous tales,
was not at all about a hated world.
It was and is a call to disregard the plumb,
as do the dancer's silken veil, the flag
folding like kelp in the just breeze,
the balloons of heat and helium.
Behold the hummingbird who will not forget
water's freedom in a falling world.
It did not have to lift toward emptiness
a moon away from its kind,
and ponder continent and season
in spun blurs eclipsed by a gloved thumb
to understand the stillness of journey.

Column. © Richard Robinson

Notes from an Urban Dig

Carolynne Baker

Fragment 1

Thang Long was once a city of lakes. Early monochrome etchings suggest that the moon was in a watery constellation when the ancients founded the city. Plans show an archipelago of lakes, canals, rivers, dikes, and bridges in a vermiculated sea of streets. As if to capitalize on the aquatic theme, even the air surrounding Thang Long is heavy with water. Every image, every vista of the city is refracted through a mist known locally as rain dust. Clear views emerge only on the condition that the rest of the city remains in metaphysical fog, momentary lapses in the fluid continuum. The water traps Thang Long into a circularity of circumstance from which it cannot escape; the city is forever caught between high humidity and rain. Occasionally, when heavy rains cause the dikes holding the adjacent river to fail, the whole city is washed away, engulfed by its recurrent theme.

When Western imperialists first colonized the city, they saw in the numerous bodies of water not a reflection of the order of things but dysentery and malaria and other unarticulated fears. In the fashion of the day, they set about sanitizing the city: channel stones were replaced with gutters; fluvial areas were dredged; the uninhabitable lakes and subterranean space were filled with sand from the riverbed. Cracked paving stones, poorly plumbed walls, miniature gates, and strange trees appeared, indistinguishable—or so it would seem—from the old dry city.

There exist tools to divine the old city. Traces of difference—a twist of Venetian must, the close gathering of watermarks, an overburdened gutter—mark the former shores. Access to these tools, however, is restricted to an elect few: a shadowy congregation of rats, waterproofing contractors, cartographers of history.

Fragment 2

Every street of Songping—even the narrowest, the most crooked, the most obscure—is lined with a tempting assortment of food: hot bread, bitter oranges, pineapples, delicate pastries filled with palm sugar and coconut, sticky rice wrapped in banana leaves, noodles flavored with lime and chili and star anise. From each of these choices, numerous other choices spring. At a certain teahouse, there are, for example, twenty-four listed permutations and combinations in which mango juice can be prepared. This figure increases considerably when the number of different ingredients that can be combined are taken into account: orange, strawberry, banana, cherry, lychee, custard apple, coconut. Even the seemingly innocuous addition of ice—or water stone, as it is known—assumes a number of forms. It may be added to the selected concoction in a great chiseled lump, or it might alternatively be shaved and stirred through with a generous dollop of condensed milk. For those in need of extra fortification, fruit liquor complicates the matter of choice considerably, although most menus fail to mention this out of respect for the local licensing laws. Songping is a city in which choice branches out from itself; even as one decision is made with the apparent effect of negating all others, options continue to appear, unexorcisable apparitions.

The availability and sheer quantity of food belies the fact that not so many decades ago, the city was stricken by a terrible famine. An estimated 1 million people starved to death even as surplus grains decayed in colonial stores, a karmic strike that spelled doom for the regime's rule. Seasoned in remembrance of a time when things were not as they are, the flavors derive a certain piquancy. The food is particularly difficult to resist.

Fragment 3

At La Thanh's heart lies an ancient quarter through which a labyrinth of streets twist and turn and crook and divide, as though laid out on the whim of a scribble. If the plan of this quarter is difficult and disorientating, it is compensated to some extent by a rigid street system that names streets by their wares as a means of keeping the chaos in check. So, for example, a great tentacled banyan tree marks the beginning of Banyan Street, and Silver Street gains its name from the multitude of silversmith shops with which it is lined. This system has the great advantage—in theory at least—of enabling street names to transcend their status as mere markers and become directions. Under this system, the city corresponds with itself. It speaks the language of its streets.

Navigating these streets, however, is not as easy as it seems. The system contains a single but nevertheless fatal quirk that compromises its supposed simplicity: the street names do not always change in accordance with the evolving streetscape. The current collection of street names relates to the streets only as they were once arranged or, rather, the street names relate to a city that never was. The city's streets stretch laterally throughout the ages, catching here, slipping there, traces of the old city randomly emerging through the new. No one can explain what governs this slippage, however, except perhaps to say that sometimes memory is more potent than actuality. Even if it was known that Peach Street was named not after mounds of the downy fruit but after the bolts of peach silk that were once dyed and traded there—this knowledge would not account for Peach Street's now being lined with spanners and gaskets. Full knowledge of the city's streets is confined to those who understand how the streetscapes are confounded by the particularities of history. Full knowledge of the city's streets is thus confined to a precious few.

Constituting a sizable proportion of the precious few are a handful of stone dragons that, having held court on the rooftops for many centuries, have silently witnessed everything. They prefer to keep their secrets to themselves, however, and divulge very little. Besides, as the dragons reach old age, their numbers are dwindling. They are becoming increasingly rare, verging on the mythical.

Fragment 4

Sometimes Dong Quan is a still city. Nightward, the city folds in on itself, wrapping its wares and color and sleeping inhabitants in a private curtain of planks and steel gates and roller doors. Corners distinguishable by day for their cairns of fruit or colored shirts merge by night into the dark tedium of concrete and steel, the duplicate materials of modernism. The noise of the city also dims as darkness falls. The incessant honking ceases. The plangent cries of the propaganda speakers fall silent. Only the mournful monotones issuing from those hawkers who have failed to sell enough during the day punctuate the calm of night.

For those who are first introduced to the city in daylight, the sleeping city seems a different place, the product of a different time. As its features relax in sleep, cartographic dyslexia overwhelms whole sections of the populace: newcomers, the inebriated, those who fear the dark.

Fragment 5

Sometimes Dong Kinh is a still city, but not very often. A number of forces act to rouse the city from its static state, the most persistent being the tropical storms that, during the city's summer months, fly in with the dusk-like dark apparitions. Warning of a storm front is first relayed by faint flickers of lightning that menace the sky. Distant thunder invariably trails a moment behind. Both elements, subtle at first, become increasingly violent as they approach the city. Thunderous detonations shake the city's roots. Great flashes of lightning ignite the sky. When the rain finally begins to fall, it rushes down in great slanting sheets. Gutters extend into the roads. Streetlights beam from the puddled ground. Impromptu waterfalls cascade from the fish-scale tiled roofs. Few unfortunates venture onto the streets under these conditions. And then, without warning, the rumblings crash off to the east, many leagues at a time. The sudden storm leaves the city spent, reflective, strangely dustless.

A second, but no less transformative, force that acts on the city is difficult to trace to a particular point in time as it—somewhat improbably perhaps—seems to occur only when no one is watching. In this temporal half-space, buildings mysteriously grow, billboards vanish, neon signs flash their pleas at the night. A bamboo scaffold, all points and spikes and bad attitude, is dismantled to reveal a facade transformed beyond recognition. A concrete slab in the process of being broken up with a hammer and chisel seems to have floated off within the hour. The urban fabric shifts as though in a dream—or a nightmare: a web of silk billowing formlessly over the city, an uncommitted lover. Even for those who accept the changes without so much as a shrug, each new incarnation makes it ever more difficult to believe in the appearance of truth as it manifested itself, that time.

Fragment 6

A strange phenomenon is at work in Dai La. With almost parasitic fervor, shanty constructions have colonized the colonial villas. Metal flanges, reinforced steel cages, rusted scraps of corrugated iron: the facades are a riot of elevational anarchy. These shanties, however, are not the result of neglect and despair. The sheer quantity of decorative objects with which they are festooned—silk lanterns, tasseled umbrellas, leafy plants in brightly glazed pots, sinuous bamboo birdcages—testifies to the fact. It is rumored that one of the most impressive collections of interwar housing in the region lies buried beneath the profusion of makeshift extensions. This is not easy to confirm, however: with each successive facade, the original facade becomes increasingly difficult to discern.

Although translating from architecture's strange tongue is a notoriously difficult enterprise, there are several hypotheses regarding the effacement of the colonial facades. For example, it could be that where the colonial buildings once defied all existing patterns of Dai La, the city is now merely reestablishing its former footprint, like a dense tropical jungle speeding to reclaim all that doesn't actively resist its advance. On the other hand, it may be that the colonial edifices have been deliberately effaced in a bid to forget the suffering caused by the colonial regime, as though their facades were inscribed with tales of starvation and incarceration and other unspeakable horrors. Nor was this suffering confined merely to the period of colonial rule: its aftershocks rippled right on through the twentieth century. The decision to split the country into two halves, for instance, was made in direct response to the ousting of the colonial regime. Although this cleaving of geography was supposed to have been temporary, it proved to have such deadly consequences that another twenty years would need to pass before the full extent of them became clear. Given that none of Dai La's leaders were ever consulted about the decision to divide the country, the shanty colonies could be read as a political protest, a bold attempt at postcolonial amnesia. And yet, conversely, it could be that these constructions merely represent the desire for additional space in one of the most densely populated areas on Earth, highly plausible in a place where people routinely sleep five to a bed and share one bathroom among forty-five. If this is the case, then these constructions signify nothing beyond themselves; all drawn meanings, all interpretations, are merely tricks of light, products of an overly active imagination. The lack of a common language between the environment and the interpreter is a source of great uncertainty; the chain of Chinese whispers complicates the simplest of translations.

What is certain is that the city is constantly at work: toiling, re-forming, molding itself according to what is demanded. It is as though the city inherently resists assuming any other form, as though its inhabitants intuit the natural order. The natural order is not as it once was, however, and under the shadow of its great washed sky, the city struggles to remember its old self, even as it cannot forget.

Fragment 7

The area that the city of Thang Long Thanh now inhabits was originally home to a number of small villages and hamlets. Although these villages and hamlets have long since merged in the process of urbanization, clues as to their former locations are contained in the temples that are now scattered throughout the city. Along with the odd ancient street name, these temples are the last remnants of Thang Long Thanh's preurban past.

Presided over by phoenix, unicorns, tortoises, and dragons—beasts thought to be capable of imparting the qualities they symbolize—the dim interiors of these temples are hung with objects that relieve the ever-deepening dark: cones of coiled incense, fat-bellied Buddhas, bronze gongs of impressive diameters. Among the forest of heavy ironwood columns hang a large number of horizontal and vertical wood panels. Deep red or dark brown, these panels are inscribed with gilt characters that appear to float in the gloom. They are the medium by which history is incorporated into the fabric of the building: their characters relate facts that constitute both temple archives and temple itself.

Although the idea of the public archive has probably nowhere else been better executed, there is one critical impediment to this public education system: the characters inscribed on these panels have long since ceased to be the characters that describe the spoken language. The once pictographic script was romanized before Thang Long Thanh was even colonized, primarily as a means by which to facilitate missionary activities. Although the spoken language remained largely intact, the visual language was completely remodeled. This broken continuity has dramatically reduced the language's power to relay information. To all but the odd scholar of the ancient script, these characters are now completely incomprehensible.

And so the histories of these buildings remain buried, mysterious, beyond the grasp of the masses. The panels now hang as little more than strange metaphysical fruits; it is possible that there is substance to their polished hieroglyphic surfaces, but it is equally impossible to prove.

Fragment 8

Where once a network of canals flowed through Dong Do, a network of roads now links various parts of the city. While it isn't precisely true to say that there are no rules governing these roads, it is nevertheless one way of explaining the anarchic state of the traffic. The line down the middle of the road commands complete disrespect. Red lights are mere suggestions. Hazard lights warn to expect the unexpected and are not lightly named—trucks reverse up freeways with hazard lights flashing, secure in the conviction that warning others of the danger renders the maneuver safe. This anarchy is compounded by the nature of the vehicles on the road. It is not unusual to see a cyclist pedaling along with a fridge strapped to the back of the bike, or five pigs, or half a curtain wall. Nor is it unusual for motorbikes to be so heavily laden with goods that there is no longer any room for the driver. In this situation, the driver will merely flatten himself over the cargo and grip the handles for dear life, feet smoking on the bitumen. This, it seems, is infinitely preferable to making an unnecessary second trip. Often slower vehicles attach themselves to faster vehicles in a bloody-minded attempt to conserve energy; pushbikes in particular can often be observed trailing in the wake of motorbikes or buses, although it is not unknown for people in wheelchairs to follow suit. In lieu of traffic lights or the concept of give way, left-hand turns are executed by sailing sideways through the onslaught. The expectation appears to be that the traffic will defy its molecular structure and flow about the vehicle as though a fluid. Miraculously, more often than not, it does. Perpetual honking is the single concession to road safety that is made. If most people manage to negotiate these roads with their lives intact, it is because nothing moves very fast, mainly because it can't.

In this city, as in countless others, there will come a time when the glee with which vehicular technology has been embraced will slow the city down. The thoroughfares will all be choked with traffic. The atmosphere will be hazy with fumes. The roads will be pocked with neglect and the vehicles all steered by wheezing asthmatics. Clairvoyance cannot control the future; it merely augurs it.

Fragment 9

Located at the confluence of two rivers and bordered to the south by a third, Bac Thanh was originally bounded by a rampart system that cleverly incorporated the three rivers and their dikes as a moat. Although the city was walled to help guard against repeated incursions from the north, the principles of geomancy were also employed to help determine the final form of the city. In this way, Bac Thanh's inhabitants aimed to insure their city against *all forms* of misfortune. The city was designed to repel all unwelcome visitors, of this world or the next: barbarian invaders, imperialist regimes, evil spirits sniffing at the city gates.

Unfortunately, the many precautions Bac Thanh's citizens took against invasion proved to be more than fully justified. An imperial city from its earliest days, Bac Thanh regularly found itself under threat or, worse, foreign rule. The city suffered much at the hands of foreign powers: one thousand years of domination here, eighty years there, battles and wars too numerous to count. Such suffering, however, merely reinforced the wall even as it highlighted its impotence. Whenever one wall was razed, a stronger wall would be erected to take its place. And so a series of walls was built to protect Bac Thanh from the outside world: bamboo walls; earthen walls; brick walls; concentric walls; hard stone and laterite walls; walls of ever diminishing diameter. No one seemed to care that the gated city performed a dual purpose: that it trapped even as it protected.

Bac Thanh is no longer encircled by a gate. The ramparts that once bounded the city have largely been demolished, and the remaining fragments stand as little more than historical curios. Nor do most people seem to wish it were any other way. Signs of welcoming the outside world are everywhere: in the looping power lines; in the great trussed sentinels; in the television antennas, the Internet cafés, the neon billboards. It is as though the gated city were a chrysalis within which an irreversible transformation had taken place. There are, however, a few who view this development with more than a degree of ambiguity. Potentially subversive impulses pass through the profusion of lines, and it can be somewhat difficult to exert control when limitation is lacking. The gated city languishes under the invisible incursion; the free city, however, is not yet fully formed.

Fragment 10

Not only is Bac Thanh no longer encircled by a gate; Bac Thanh is no longer Bac Thanh. Nor was it originally Bac Thanh, but a small village known as Songping. Over time, Songping developed into the township of La Thanh, which in turn came to be known as Dai La. When King Ly Thai To transferred the capital to Dai La in the Year Canh Tuat (1010), Dai La came to be known as the citadel city Thang Long. After a long period as Thang Long, the city then spent ten years as Dong Do, several decades as Dong Quan, and several centuries as Dong Kinh before it was ever known as Bac Thanh. It is therefore slightly misleading to say that a series of walls protected the one city. Perhaps it is more correct to say that a series of walls protected—or failed to protect—a number of different cities; or even that the city is less one city or a series of different cities as a reincarnated city: chronologically blurred, thematically repetitious, forever bearing the sacrifices of its former selves.

During the second half of the twentieth century, the incarnation in which Bac Thanh gained global fame was as Hanoi. The name became metonymic, not only with the northern half of a country that had split in two but also with one side of an international war that flashed into living rooms around the world on a nightly basis. During this war, it didn't matter that Hanoi's gates had not survived colonization. No gate could have protected against the bombs that rained down on the city. Yet even as tracts of Hanoi were reduced to rubble, there were stirrings in the ash. Another Hanoi was set to emerge.

Are these two Hanois the same city? Or are they separate entities that have chanced upon the same name? The 50 percent of the population who were born after the war don their Levis and Nikes and simply shrug at these questions. What difference would it make either way?

Fragment 11

In Hanoi, it is a well-known fact that those in the afterworld need to be nourished and loved and entertained; the needs of the dead mirror almost exactly those of the living. Ancestral altars, found at the highest point of almost every house, reflect the care that is taken to sustain the spiritual world. Offerings of fruit, incense, and flowers weigh down the most modest ancestral table. Sometimes the fruit is omitted for a quart of whiskey, a bowl of presmoked cigarettes, and a pack of cards. As the happiness of the recipient is a high priority, it is important that the offerings are selected to reflect their tastes. Anyway, these habits are far less damaging to the dead.

One of the great sorrows of war is that soldiers who die in battle have no designated point at which this exchange of provisions can be made. It is a great fear that these souls are destitute despite the fact that their living provide handsomely for them. And so, on the fifteenth day of the seventh moon by the lunar calendar, Hanoi's inhabitants throw rice, popcorn, and salt around their houses so that should a lost and hungry spirit happen to wander by, he or she might be sustained on the journey home. Since it is impossible to know when, or even if, a spirit arrives home, this offering of food to the night continues long after truce has been declared. It is best to cover all contingencies.

Fragment 12

Unlike the deviant concubines who were condemned to atone for their sins by spending their lives weaving a fine white silk on an island in one of Hanoi's many lakes, Hanoi does not weave symbols of purity to right its wrongs. Issues of morality and redemption are beyond the city's concern. Rather, Hanoi accommodates all its threads, its moods, its facets, and gathers them into a fabric in which competing concerns coexist even as they pull on one another. Change, continuity, regression, linearity, circularity, recurrence, appropriation, rejection, rationality, order, chaos: all compete for articulation in the weft and warp. Where conflict occurs, the weave coarsens. Where an overriding narrative breaks through, the fabric is smooth. The idiosyncrasy of weave renders the fabric patchy, but it is unique and irreplaceable nevertheless.

For those who know how to read the fabric, the city's evolution can be picked out in the arabesques, the flutings, the tracery. But what of the moth holes, the frayings and ravelings, the places where it is possible to poke a finger through the thinning weave? The uncharted is arguably the most crucial constituent of truth, but it is solemnly sworn against disclosure.

IV

Getting to the Future

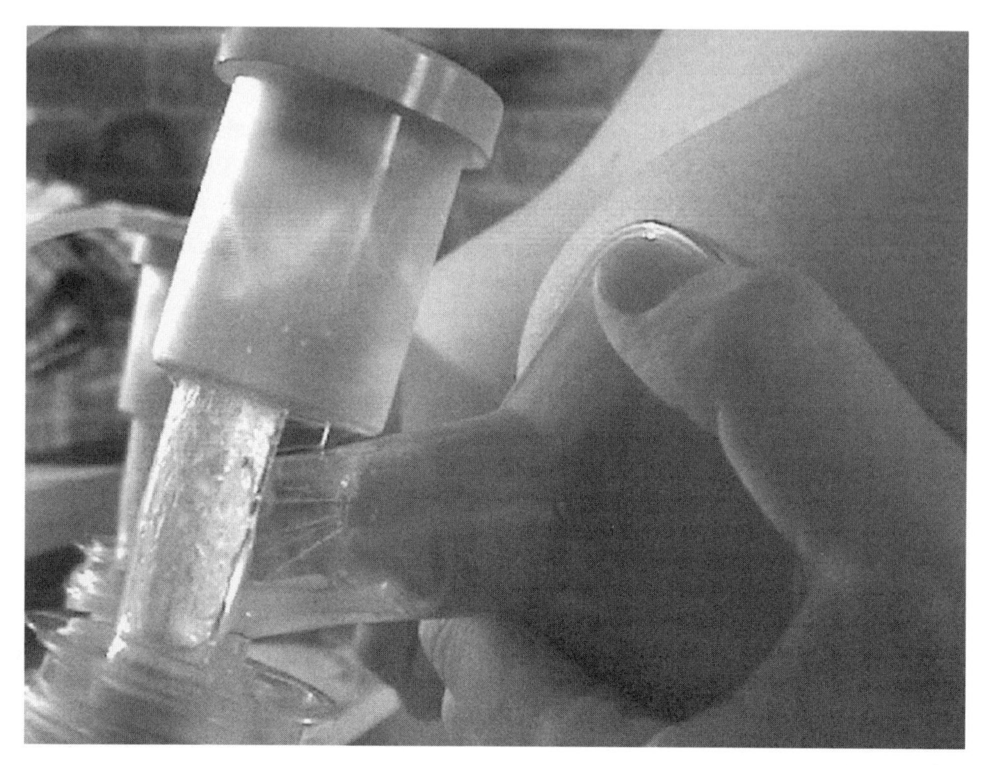

Monika Wuhrer, *Visceral Circle*

Quest for a Future Perfect

Kathleen Creed Page

> *Bringing the body and its vital energy up to speed with the age of instant teletechnology means simultaneously abolishing the classic distinction between* internal *and* external.
> —Paul Virilio

Life by definition is change—random combinations followed by a selection of forms most fit for the environment. Being a "good fit," of course, is dependent on the prevailing conditions, and adaptability is the human animal's most successful trait. Historically, we have understood evolution to be the process of natural selection; however, as this century turns, we are constructing and deconstructing our bodies using a set of self-fashioned selection tools. We have entered an era where chimeras of information, born of biotechnology and microelectronics, breach the boundaries that once existed between human, animal, and machine. This has prompted us to consider the notion that we are fabricated beings struggling in a network of high-tech culture. From this perspective, we can anticipate unique couplings with our cyborg reality and a disruption of our more persistent dualities: self/other, mind/body, culture/nature, reality/appearance, whole/part. In our current biotechnological state, it appears that these oppositions have been torn apart and that the story is being retold. On the horizon, the sun still rises and we awaken with some certainties, but we are also experiencing a phase shift—a turn in the narrative that occurs so sharply no character remains the same. Our biotechnological practices have provided a new set of terms that have begun to reshape our sense of self. Terms such as *donor eggs, donor sperm, in vitro ovaries,* and *host wombs* are becoming instrumental to narratives of reproduction that have turned sharply away from the body and veered deeply into the images surrounding "body parts" or "partial bodies." Doctors refer to women as *surrogate uteruses;* judges address women as *surrogate reproduction vehicles,* and ethicists report that fertile women function as *therapeutic modalities* for the infertile.

These terms fracture the traditional idea of the maternal and diffuse into the currents of identity formed by a more instinctual mode commonly referred to as our second nature. Beyond this lies an incipient *third nature,* a state of being derived solely from our own technological intervention. This evolution of our bodily state of affairs is analogous to the changes in the boundaries of place recognized by McKenzie Wark, author of *Virtual Geography: Living with Global Media Events.* He suggests that information technology has introduced a "qualitative change in the social relations of culture . . . [where] information moves faster and more freely than people or things," and that the geography of our second nature has been overlaid by "a third nature of information flows, creating an information landscape which almost entirely covers the old territories."[1]

Our new reproductive technologies have become carriers of this third nature as new terms begin to stir our imaginations, and we create a language that promotes a reproductive climate where gamete exchange and embryo commerce "cover the old territories." This biotechnological revolution and the exchange of body parts has not arrived de novo. Desire to access human nature and manipulate reproduction has existed for thousands of years. Magical rituals, coitus interruptus, extension of lactation, abortion, and infanticide have long been practiced as options for reproductive control, and their outcomes have become factors in a selection process. Today, if the desire to control human nature were taken to its limit, the biomedical establishment would have us believe that the perfect human being would be attainable through design and implementation—a utopian ideal.

Growth in genetic theory aided and influenced new developments in reproduction. In 1936, the influential geneticist and Nobel laureate Herman J. Muller published *Out of the Night,* a book that envisioned a comprehensive biological future divorced from state control. His notion was that a more favorable reproductive strategy would be developed within a "cooperative structuring of society" and that any eugenic conception devised within capitalist circumstances was doomed.[2] Muller maintained a strong socialist conviction that supported his aims for political reform. For example, his strategy included a change in the economic position of women by making contraception widely available, thus freeing women from sole responsibility for child care. Years ahead of his time, Muller recognized that traditional eugenic practices, such as sterilizing carriers of deleterious genes, were hopeless. Instead, he hoped that future technologies would be developed to ensure our ability to carry out a direct selection process. These approaches would include "transplantation of ripe or fertilized egg cells from one woman to another, the development of egg cells without fertilization, and the manipulation of embryos, in particular artificial insemination in connection with the selection of the desired genetic material."[3] In fact, Muller claimed that germinal choice "turns out on closer inspection to be the most practical, effective, and satisfying means of genetic therapy."[4] Although Muller originally aspired to a rational volunteerism regarding reproductive choices, he eventually took a more pragmatic approach: technology.

After this new fix had been sighted, the focus moved to a more modern arena, where the language of the biotechnological enterprise began to dictate the content and structure of human values. Linguistic choices became the currency for reordering our images. Even the phrase *reproductive technology* implies the mechanization of parts. Gametes have become more easily recognized as reproductive units or replaceable objects that can be cataloged and parceled out, a utilitarian approach to taming passion and undermining intimacy.

To engage with this new form of knowledge places us in a field where language is suspended in a thin medium of information flux and flow. It is here that the new "repro-tech" proves to be a source of abstraction and alienation—encoding a woman's body in the vast rearrangement of language and metaphor. How can we understand these new forms? Perhaps we must look at where our thoughts and actions intersect, where wombs and embryos emerge as partial bodies, migrating among domains, homes, and languages. In this expanded field resides a shifting sense of value generated by novel reproductive practices.

Is it meaningful to imagine free choice within the context of the current sociopolitical arena? Is seeking fertility an individual choice, or is it coerced from our social and historical climate? Perhaps the problem of infertility is a social construction, and the necessity for intervention is created only within a context that defines human nature as a subject for technological design. Ultimately the need for the female body to carry the fetus will be averted as artificial wombs are created and implemented as just another option. Today's reproductive practices have driven a wedge into our sense of connection that urges us to question what it means to be human. It is even possible that we are no longer users of nature but producers of nature, an inversion that may result in an environment completely derived from human artifice. The reality of these projections may be observed in the reverberations of the twentieth-century quest for a "future perfect." Muller was not so far off imagining that the most efficient future involved sperm and egg cell manipulation, a breeding strategy more suitable for ensuring selection.

Current reproductive choice allows our rational power to completely dominate our bodily instincts. This desire for control has promoted an almost unquestioned acceptance of extracorporeal reproduction—our cyborg destiny borne of a third nature. Under this condition, one forgets that the need to control nature in the interest of human freedom carries with it the need to acknowledge the bodily dimension of human existence, the compulsions and passions driving our unique sense of necessity.

It is also possible that the technological interventions leading to the re-creation of human being as a techno-organic hybrid have lead to a genetic signature of sameness. As we become increasingly driven by a desire for more perfect outcomes, perhaps we will eventually find difference unacceptable and grow intolerant of error. It is evident that the fertility marketplace has invested heavily in sperm-egg exchange and embryo commerce. But what about the shared space between two bodies, two pulses, and their transformations?

The new boundaries generated by repro-tech constitute a form of discourse outside maternal language. Discussions range from the use of donated eggs for infertile surrogate mothers, to the notion of genetically modified "designer babies," a strange image that spins off recklessly from the current embryo debate. Perhaps we should clarify the new techno-image in terms of a woman engaging with assisted reproductive therapy, or ART. First, she is encouraged to break through what appears to be an arbitrarily imposed limit to her reproductive potential, her biologically determined state. After she has been shown to bear the mark of infertility, she becomes an exercise in ART. Her blocked fallopian tubes become a bypassed body part, and her ovaries surface as central. Imagine the physician, with his laser optics, peering into the abdomen of this woman:

All is quiet beneath the ovarian field. The new crop of shimmering eggs have become grotesquely enlarged within the dark interior. Subliminal messages arrive like a thousand black birds taking flight, each molecular flutter agitating the pulse of a delicate internal rhythm. Hormones, designed to hit their targets compress the ebb and flow of the women's cycle, while other circulating drugs suppress the dominant egg cell from ripening. Instead, a flurry of gametes surge toward rupture, super ovulation occurs, and the doctor beams as he observes the Herculian release of a dozen eggs. The woman's distended abdomen winces as the eerie eye of his laparoscope glares through the dense fluids. The oocyte retrieval has begun.

Oocyte retrieval, a procedure that extracts the egg cell from the woman's ovaries, has been revised recently, and the reproductive technologists claim that this procedure is "practically painless." But others have found that there are some layers of the body one cannot numb. Compared to earlier techniques, these ultrasound-guided follicle aspirations are less productive in terms of harvesting the eggs; however, they are safer, require less equipment, and are lower in cost. With these improvements, the technology has become more available to patients, and health care insurance often reimburses the service.

However, let us not begin legitimizing technical terms simply through their usage. A different question needs to be asked: What has happened to the body of this woman? She has been treated with an array of steroids, her ovaries have been hyperstimulated, and an abnormal number of egg cells have burst from the ovary. In addition to oocyte retrieval and the associated discomfort during the egg harvest, she must also suffer the anxiety of waiting as the collected eggs, gathered in sterile nutrient broth, are brought together with ejaculated sperm in order to coax their fusion. In the end, this in vitro fertilization (IVF) results in an egg-sperm union outside the human body.

Again, the woman's body parts are mere shadows of sound. The organic matter, in silence, reflects the voice of the unborn as the newly fertilized egg is delivered back into her body and falls freely within the fallopian fluid soon to flow to the uterus and take root.

The technique was originally "designed" to bypass sperm motility problems by positioning the already fertilized egg back into the woman's womb. Imagine a beginning such as this—the doctor preparing the impregnation as he contemplates the inner sanctum of the anesthetized woman.

It is clear that the naming and implementing of these procedures allows us to imagine a new form of reproduction. Body parts, such as spare eggs and sperm as well as extra embryos, are currently being frozen for future use, and with this arrives a new set of terms: *egg banks, sperm banks,* and *embryo banks.* Of course, these banks are quality controlled, employing the most sophisticated tools to detect genetic defects and repair embryos tagged for implantation. Other options include conceptus transfer, more often described as the insemination of a fertile woman with sperm from the infertile woman's partner. This illicit "coupling" would be followed by a flushing of the embryo and a subsequent transfer to the infertile woman's womb. Women may even become professional donors, thus washing out their embryos every month. Perhaps these embryos would then be frozen, shipped out from conceptus transfer centers, and allocated according to a computerized central banking system.

These new practices, as well as artificial wombs, gestation of human embryos in other species, nucleus substitutions, genetic engineering, and cloning, provide the foundation for a cyborg state. How should these new optional modes of reproduction be evaluated in terms of participatory politics—informed public consent based on a self-determined choice? Is there any reason to reconsider the manipulation of these body parts: the egg, the sperm, the embryo? Are these body parts or are they partial bodies, since upon further development, a potential *other* will be generated? What principles govern the placement of this partial body? What will be the impact on our relational values if new forms of discourse move the physical concept of mother outside the female body? How will the human being of the future differentiate itself from isolated or scattered body parts on one end of the continuum and experimental hybrids on the other? As the distinguished geneticist Joshua Lederberg claims, "Human legal privileges will be granted to those objects which are similar enough to human beings in appearance to move human conscience, and which are not too expensive to keep."[5]

Our exploration of the "real" and the "possible" outcomes of this new practice are an action for which we are in some sense unprepared. The material codes of our existence can now be used to shape evolution directly. Perhaps human nature has entered an era of re-creation and has begun to view itself as a partial body, possibly as an entity *outside of itself.* It has become clear that the sexual act, once prerequisite for human reproduction, has been severed from conception. In vitro techniques have displaced the fertilization process to an extracorporeal landscape or, should we just say, a sterile glass dish.

What is of concern here is the level of reproductive manipulation. Technology is no longer a means for recognizing already present needs, but in itself makes options available that influence our choices and actions. We deal in body parts, using their implementation as new avenues to new ends. Although these techniques appear to have arisen in response to the need of infertile couples, another reading suggests that biomedical research may be driven by the desire to use the embryo as an experimental subject. Thus, genetic manipulation would be more powerfully rendered and human genetics would be more accessible to intervention. Biological enhancement would no longer be subject to chance and necessity, and the boundaries of human origins would be inevitably ruptured.

Ultimately, our taming of the life force must remain in balance with the autonomy of the life force. To be subject to something unpredictable and mysterious provides a respite from a human-centered world and enhances our sense of otherness. It is possible that unlimited use of reproductive technologies may weaken our sensibilities toward the plurality and unpredictability that marks life, while our capacities to distinguish between nature and human artifice will dull as we begin to dictate how the new will look.

Now is not a time to simplify the technopolitics of reproduction into senseless contradictions, spinning the rudder of the sociopolitical movement into a downward spiral. It is a moment of acknowledgement—a pause in which to recognize that alternative methods of conception have diversified historical practices into a new reality. It is a harbinger for the arrival of a cyborg identity—a concept that challenges our assumptions of an essential self. The very potential that could drive the imagination toward a more creative sensibility of other is countered by the outcome of technology, a product line that propagates sameness through cloning.

We may now represent the genetic mother, the physiological birth mother, and the actual caregiving mother in a signifying chain that opens heretofore uncharted domains of thought. Surrogate mothers have become surrogate wombs, reproductive vehicles, or therapeutic modalities; new life has become detached and transported to foreign destinations. Perhaps thoughts about surrogate wombs and embryo commerce are simply junctures in reconceptualizing maternity. We might simply accept the technolanguage of reproduction that privileges the sperm and egg, and promote the image of woman as womb—as site for exchange. Actually, there is nothing new here in terms of women's historical positioning. Or is there? At the moment, all we can really determine is that the linguistic field that figures women as shadowed body parts is currently being sown with new seeds, possibly in a womb for rent.

In essence, the implementation of technology has begun to move reproduction beyond chance operations; however, in our attempt to reduce the impact of uncertainty, we may also reduce the sovereignty of human life at its most elemental level. The steps we have taken in the reproductive sphere have driven us closer to claiming our position as producers rather than as participants in a compelling drama governed by nature. But our ability to initiate thought and action rests squarely on the evolution of meaning, and one of the critical questions embedded in that meaning is, "What is human?" Evolution is dependent on an openness to unpredictable outcomes. By excluding chance from human reproduction, we risk banishing what is essential to the human experiment and put humanity out of touch with the play of possibility. Perhaps, in the end, the elusive element of uncertainty restores the impulse of promise and desire that awakens in the unprecedented fact of our birth.

Notes

1. McKenzie Wark, "Antipodality," in "Intellectuals and Global Culture," eds. C. Blake and L. Blake, special issue, *Angelaki* 2, no. 3 (1997): 22.
2. H. J. Muller, *Out of the Night: A Biologist's View of the Future* (London: Gollancz, 1936), 51.
3. Ibid., 135.
4. Ibid., 258.
5. J. Lederberg, "Orthobiosis: The Perfection of Man," in A. Tiselieus (ed.), *The Place of Values in a World of Facts* (Stockholm: Almquist and Wiksell, 1970), 49.

Intelligent Robots or Cyborgs

Kevin Warwick

I've got a chip on my shoulder. Well that's what some people say. In fact, it's not far from the truth. On the morning of March 14, 2002, I found myself lying on the operating table in Theatre 1 at the Radcliffe Infirmary, Oxford. I was ready to have a small array of one hundred electrodes fired into the main nervous system in my left arm. I knew that once the array was in place, we would head back to my lab at Reading University and try to link my nervous system directly to a computer. It was an operation I didn't need for medical reasons but purely for research—in order to find out.

The operation had been going for about an hour and a half before the neurosurgeons were ready to fire into my nervous system. Peter Teddy, the consultant neurosurgeon, called for a large microscope, which was wheeled forward and rotated into position over my arm. Out of the corner of my eye, one of the TV monitors flicked through channels, displaying the view through the microscope. If I had turned my head slightly, I could have seen the monitor clearly, but I wasn't sure that I wanted to. Mark Gasson, one of my researchers, said that you could clearly see both the array and the exposed nerve fibers and they looked just as we had expected. Carefully and painstakingly, Peter maneuvered the array head into position over the nerves.

Peter doused my nerves with more local anesthetic. I could feel it swishing around as it touched parts of my arm that were not fully anesthetized until then. Next, the impactor unit (a fancy name for a pneumatic hammer) was switched on and brought into position. The unit was charged with compressed air, the head of the impactor sucked up and then forced down, just like a hammer. The head had to hit the array in order to force the electrodes into my nerves, like hammering a nail into a piece of wood.

"Kevin, can you wiggle your fingers?" asked Peter. I tried and found that I could. "Right, then. Are you ready to go for it?" As Peter asked the second question, I knew this could easily be the last time I would be able to move or feel some, or all, of my fingers. But this was what I was here for. "Let's go for it," I replied. Peter nodded. There was a sense of hush and expectancy in the operating theater. All eyes, except my own, were on either the impactor or the TV monitor.

"Okay. Here we go" said Peter. I heard a loud click as the impactor fired and felt a ping of electricity in my thumb, but nothing too much. Peter announced that the array had not gone in. "Let's try again," he said. Peter checked that I was ready and fired the impactor for a second time. Once more, it failed. Peter asked me to move my fingers, which I did. Everything was still okay. He tried once more, but to no avail.

The equipment was checked for leaky pipes, but everything seemed to be in order. Peter tried again, but it still didn't work, and once more, but no luck. All in all, Peter tried seven or eight times to force in the array without success. There was nothing I could do but feel disappointed, tremendously sad. We had come so far. It had taken so much time and effort and people to get everything together, and here we were, falling at the final hurdle. All that was left was to pull out the array, stitch me up, and we might as well go home.

But then it was spotted that the two connecting tubes on the impactor unit had been connected wrong, so that the unit had been sucking when it should have been blowing. It was merely dropping onto the array head by means of gravity. The effect was about the same as trying to crack open a coconut with a teaspoon, so the connecting tubes were switched over.

"Okay, Kevin, are you ready?" asked Peter once more. "Go for it," I replied. Peter then fired the impactor. This time the click was much louder, and I felt a bolt of electricity down the inside of my thumb—zap! Peter checked once more that I could move my fingers and had not lost any feeling and announced that the array had gone in successfully.

Relief poured out from everyone there. To make sure that we really had a connection, Peter used the impactor to hit the array twice more. Each time I felt a large zap on the inside of my thumb. At last, the array was sitting comfortably, and the incision could be closed up.

All of this might seem to be rather a lot to go through for a scientific experiment. Taking on such dangers is not the usual fare for a university professor. So what was it that made me do it? What was I hoping to achieve? For the answer to that, you have to step back and look at my history with robots, in particular investigating the intelligence of robots—how it compares with that of humans, what this experiment means as far as human existence is concerned, why it's important.

Humans are certainly good at doing one thing, and that is being human. In our present *Homo sapiens* form, we have been around for 100,000 years, not a significantly long time in comparison to planet earth or indeed many of the other living creatures. One thing appears to be clear, and that is that evolution keeps on going, and things change. Over a period of time, creatures adapt. Either they become more successful and survive, or they die out. History is littered with species of one type or another that were once extremely powerful but are with us no longer. As far as humans go the big question is: What of our future? Will we adapt, survive, and retain our position, or is it likely that something more powerful than humans will emerge to steal the spotlight?

If we compare human abilities, in particular our intelligence, with other creatures on earth, on the whole we come out ahead. As a result, humans have, without doubt, enjoyed a wonderful period of relative dominance over other species. We have used our position to treat many other creatures with utter contempt. In some cases, we merely destroy the habitat of other living creatures; others we keep captive. The most "fortunate" we farm, killing them for food, for political reasons, or, unfortunately on many occasions, just for fun. Indeed, ending the lives of creatures less powerful than ourselves is for some the social norm.

As far as we know, humans are the only creatures who have explored and can exist in virtually all regions of the world. In the past fifty years, we have even continued our explorations beyond our own planet. So what is it about humans that make this domination possible? What is it that we have that makes us "better" than other creatures? Many creatures exist that are, without assistance, faster or stronger than humans. So how have we gained our position? The answer has to be our intelligence, in all its intricacies.

In many cases, physically superior people are controlled by others, possibly because they are intellectually more capable. Machines, such as robots, have been created to do things that humans either do not want to do or because they can outperform humans due to their strength, speed, accuracy, and reliability. This works well when the machine is programmed and merely carries out the tasks—in particular, when it has little or nothing that could be construed as intelligence.

Humans in many parts of the world have in fact become dependent on machines of one type or another. Most of us could not imagine life without a phone, washing machine, or automobile.

For many in the Western world, the same is now true of credit cards and the Internet. Today we live in a machine-based technological world. We rely on machines for our way of life; we trust machines; we ask them questions and rely on their answers. We have grown to expect a certain standard of life that can be attained only with their help. Human progress is seen in terms of what new machines can do for us now or what more they will enable us to do in the future. Not only do we physically exist alongside the machines, but our bodies have also been tuned to rely on the standards that are provided by technology. The food we eat, the water we drink, and, in many cases, even the air we breathe depends on machines for the quality provided. Our bodies are delicately adapted to deal with the bacteria, germs, and microbes present in the technological world of today.

It is a basic rule of evolution that species must adapt as the environment changes if they are going to survive. But through technology, humans have, to a certain extent, modified the environment as needed. Each machine usually needs someone to decide when and how it should operate. A human is needed to switch the machine on. Once it is switched on, we rely on the machine to go through its programmed sequence of events in order to achieve the desired goal. Once the steering wheel of an automobile is turned, it is expected that the automobile itself will turn in that direction. In each case, we expect the machine to carry out a specific task once we have told it to do so, with us calling the shots.

In recent years, however, robots and computers are being used to make the decisions as to whether another machine should carry out certain actions. The decision might be quite a simple one, for example, switching the heating on in a house, or as complex as buying stocks. Lately machines have been employed not only because of their physical capabilities but also so they can handle our decision making. As time passes, machines are being allowed to take on more roles and this necessarily leads us to ask, How far will it go?

Nowadays, machines such as robots with computers for brains raise serious questions as to whether humans will continue to control the future. If a robot can outthink a human—and is allowed to act on these thoughts—then a rather different picture emerges. Importantly, because human intelligence is key to our status on earth, we must ask, Is it possible for robots to be more intelligent than humans? If the answer is yes, then potentially we have a serious problem on our hands. Would intelligent robots be content with us, less intelligent beings, telling them what to do, particularly when they would know better? While it is fine to have machines that are more physically capable than humans, it is quite another thing altogether for them to be more intelligent. Humans have evolved in a natural world and have now created—in the West, at least—a technological world in which we operate and on which our existence depends. We have a cozy relationship with the machines of today. They are extremely helpful, take on many of the burdens of everyday life, and allow us to do things that we would otherwise find impossible, like flying. But the performance measures and the intelligence required in today's technological world are very different from those in the natural world that existed several thousand years ago.

Over the next few years, due to the relatively slow process of biological evolution, human brains are likely to stay roughly the same size, with roughly the same number of brain cells that they possess now. Perhaps on average, there will be a slight increase, but this will be relatively insignificant. Even the most conservative estimates indicate that it will not be long before stand-alone computers outpower the human brain. Rodney Brooks of MIT said recently that "the amount of computational power in a personal computer will surpass that in a human brain sometime in the next twenty years."[1] Meanwhile, Ian Pearson of British Telecom Labs predicts that it is more likely to be within ten years. It must also be remembered that computers do not usually tend to operate on their own; they are almost invariably networked. What, then, is the brain power of a network? There can be little doubt that it will not be too long before, in terms of share brain power, computers will outperform humans.

It is therefore important to consider which aspects of intelligence should really be considered useful. If we were to choose those aspects, which were important for life and human domination over other species, if we knew what they were, then what would happen if a machine could surpass humans in those areas? Would a machine not at least question why it was doing certain things for a human, a less intelligent being?

It could be argued that no matter how intelligent machines get, as long as there is a human who can switch it on or off, humans will stay in control. But think again. Even now, would it be possible to switch off the Internet completely? We might be able to disconnect a few computers, but so many people depend on it for their daily existence that as a practical reality, it is simply not possible to turn it off.

In truth, it is difficult to conceive realistically of any important aspect of human intelligence in which robots will not be able to exceed in the near future. At this point you may be saying, "Yes, but robots will never be . . ."—filling in the gap with words like *conscious, self-aware, creative,* or perhaps *emotional.* What you actually mean is that they may not exhibit such characteristics in the same way that a human does. It is incorrect to conclude that because robots are unlikely to be approximately equivalent to humans, they will always be subservient to us. The fact is that we already have robots that in their own way exhibit these characteristics themselves—not human self-awareness but robot self-awareness, for example.

It is because robots are different, because they have distinct advantages, many of which we know about already, that they can be better than we are. In this way, robots can dominate humans physically through their superior intelligence. What matters clearly is performance. For some time, we have been able to witness computers performing feats that we consider important aspects of intelligence, such as mathematical equations or fact retrieval, easily outperforming humans in doing so. We might then say, *Well, if a computer can beat a human, it can't be such an intelligent act after all.* Do we keep making that excuse until we have run out of intelligent acts?

So is that it? Will it be the case that in the next twenty or thirty years, robots will take over and probably treat us just as we treat less intelligent creatures now? Will we face a future in which robots keep humans on farms, keep humans as pets, or simply kill us for fun on a nice sporting robot weekend? Or is there an alternative somewhere ahead?

Humans have, until now, been fairly successful in evolutionary terms. We have to be honest, though: we are very restricted in what we can do. Obviously, we have distinct physical limitations; that is why technology has been employed to help us. But we are also restricted in how we sense the world, with only five basic senses at our disposal. Technology has been used to help us out thus far, by converting information in the ultraviolet, infrared and X-ray spectra into visual forms that our brains can understand.

But there are two factors that are perhaps far more important than the way in which we sense the world. First, human brains have evolved to perceive and understand the world in terms of three dimensions—four if you include time as a dimension. This suffocates human thinking and beliefs about what is possible and what is not. Meanwhile, machines have the ability to process multidimensional information; they have the potential to perceive the world in hundreds of dimensions.

In fact, we now have enormous problems in attempting to deal with the plethora of information that technology throws up. Even greater than this (and we must be completely honest on this one), in comparison with technology, the way in which humans communicate is so poor that it is embarrassing. Human speech is serial, error prone, and an incredibly slow way of communicating because of our use of mechanical, low-speed sound waves. Our coding procedures, called languages, severely restrict our intellectual abilities as all other thoughts and ideas must be transformed into speech signals that do not accurately represent the original concept.

So let us look at a possible alternative future scenario to one in which intelligent robots rule the planet. Is it possible for us to conceive of a creature that starts out as a human but takes on technological capabilities in order to upgrade their potential? In science-fiction terms, such creatures are referred to as cyborgs—cybernetic organisms, part human and part machine. Would it be possible for such a creature to understand the world in many more dimensions, to have the memory and mathematical capabilities of a computer, to sense the world in all sorts of ways, to have physical abilities much greater than any human and, most important of all, to communicate not by means of speech signals but by thought signals alone? This truly would be a more natural form of evolution. We can look at a future in which it is not intelligent robots that dominate the planet or humans but rather cyborgs acting as upgraded humans. How is this to be achieved? Quite simply, with silicon chip technology implanted into the human body—merely one or two very small devices that connect directly with the human nervous system and brain. In this way, electronic signals in silicon become electrochemical signals in the human body, and vice versa. Ultimately, no doubt, the device will be implanted by means of a swift and simple injection. At first, microscale devices will be employable, but if the technology develops as expected, this will probably be surpassed by nanoscale connections. What we are looking at here is not extra memory being connected directly to the brain but implants used as interfaces linking the human nervous system to a remote computer network by radio.

Humans will need to be educated in the ways in which it is best to use this technology. In the fullness of time, our children's children will look back with wonder at how their ancestors could have been so primitive as to communicate by means of silly little noises called speech.

With brains linked to technology, there will be no need to learn much. Why should we, when the computer can do so much better? There will be no need to remember anything because computers have better storage facilities. When it is required to recall something, then it will be possible simply to download the necessary piece of information. It will even be possible to relive memories that you didn't yourself experience in the first place.

Perhaps you are thinking, *Well this is all well and good, but it is not really going to happen; it is merely from the pages of science fiction, a flight of fancy.* Well, think again. It was the first step on this path that I took in my operation of March 14. As a result, by means of a radio transmitter/receiver unit for a period of over three months, my nervous system was linked directly with a computer on a regular basis. We were able to transmit signals from my nervous system. During that time, we carried out a range of experiments.

Signals from the nervous system in my left arm were used to control the movement of a remote robot hand. As I opened and closed my own hand, the neural signals from my brain that carried out the action were also used to control the robot hand. My body was effectively extended with the computer to include a robot hand.

Next, my nervous system signals were used to drive around a small-wheeled robot: left, right, forward, backward.

The same signals were then used to decide on the mood of a group of small robots called the diddybots. With my hand open, the diddybots acted in a friendly fashion, flocking together. With my hand closed, the diddybots tried to get away from each other as fast as they could.

Using my own neural signals, I could control the movement of a virtual me inside a virtual house. I could move the virtual me from room to room. Inside each room, I could select objects on a menu and choose to operate specific pieces of technology. I could switch on lights or a coffee maker. But when I did, actual lights or a real coffee maker was activated. I could control my local environment merely by signals from my nervous system.

Messages from my nervous system were used to control the appearance of jewelery worn by my wife. As I squeezed her hand, the jewelery turned from bright red to a luminous blue. Ultimately it will be moods and feelings that direct the jewelery.

With my hair shaved off and electrodes positioned around my head, we were able to link and associate signals picked up in my brain with those being witnessed in my arm. I wore a blindfold, and the output from ultrasonic sensors fitted to my cap was fed down directly onto my nervous system. As I moved toward an object, I felt more and more pulses of current on my nervous system. As I moved away from the object, the pulses died down. In this way, I could move around and avoid objects without any visual input. I had a sixth sense, a batlike sense, with signals traveling directly from the ultrasonic sensors, along my nervous system to my brain.

On May 20, 2002, I traveled to Columbia University in New York, and from there, an Internet link was created between my nervous system in New York and my laboratory in Reading University in the U.K. I was able to move the robot hand around again using neural signals via the Internet. This time, I was in New York but the robot hand was in the U.K. As a cyborg, my nervous system extended across the Internet. We were also able to send pulses of current from Reading University via the Internet to New York. When three pulses were sent, my fingers were stimulated three times.

Back in the U.K. a few weeks later, touch sensors on the fingers of the robot hand were employed to provide signals directly onto my nervous system, allowing me to detect just how much force the hand was applying. With a blindfold on, I could get the robot hand to grip an object with just sufficient force. By employing the same method I had used earlier to move around a small wheeled robot, using signals from my nervous system I was able to drive myself around on a wheelchair: forward, backward, left, right. I wasn't a good driver. In the final experiment, my wife, Irena, had electrodes pushed into her median nerve as well. Signals were picked up from her nervous system, transmitted across the Internet, and played down onto my nervous system. The same thing happened in reverse. When she opened and closed her hand four times, I felt four pulses on my nervous system. We had successfully achieved, in a very basic way, the first direct nervous-system-to-nervous-system communication. Obviously, with implants positioned directly in the brain, rather than merely the nervous system, this will result in a primary form of thought communication.

Far from being science fiction, all these things are science fact; they have been achieved in practice. Quite clearly, the next few years will see many more such trials. Brain implants linking memory and motor functions of a human brain to a computer are immediate steps. Now that a cyborg path has been successfully trodden, the route to a cyborg society is definitely on the horizon. In this way, humans will be able to evolve by harnessing the superintelligence and extra abilities offered by the machines of the future. It will not be a case of robots acting against humans but rather one in which we join with them. It will not be necessary for everyone to become a cyborg. There may be those who would prefer to remain a human.

If you are happy with your lot as a human, then so be it; it's your choice. But remember this: just as us humans split from our chimpanzee cousins many years ago, so cyborgs will split from humans. Those who elect to remain human will become a mere subspecies.

With extrasensory abilities, a high-performance means of communication and the best of human and machine brains it will be cyborgs that take control. For me, there is no argument. My goal is clear: I want to be a cyborg.

Note

1. Rodney Brooks, *Flesh and Machines: How Robots Will Change Us* (New York: Pantheon Books, 2002), 197.

Lucila Wroblewski, *Silent Landscapes Second Series-Gisela #13*

De-evolution and Transhumanism

Joan Maloof

> *Everything up to the time of the Earth was brought about by divine generations of the spirit; that which is to follow will be brought about by man himself.*
> —Rudolf Steiner, 1923

Here's the most interesting thing about evolution: as soon as we understood it, we began to change it.

Thanks to the diligent work of fossil hunters, carbon daters, and the like, we can now look back in time and tell the remarkable Evolution Story. The delightful tale of stardust-to-bacteria-to-fish-to-birds-to-mammals-to-humans is told with a religious reverence. This is why we have evolved, say the priests and priestesses of the Evolution Story: to be the sense organs of the universe, to see and hear and tell of this great happening. And, indeed, we are the only creatures who comprehend the mechanics of evolution, and, yes, it is miraculous.

Besides the great bacteria-to-mammals story, there are a few other things that biologists know about this force called evolution—at least the way evolution has worked up until now. First, evolution "likes" many different types of living things. As evolution progresses, more creatures are invented. True, some of them die out, but overall the trend is toward an eclectic variety of shapes, colors, and sizes. The number of different organisms that exist on the Earth now, in this geological age, is greater than the number that existed at any time in the past. Small boys, in particular, may feel that they live on a depauperate planet, as we no longer have dinosaurs, but you can reassure them that we have more species now than we had in the age of the dinosaurs. Evolution tends toward variety, and given enough time will create it. Second, this trend toward more, more, more, is occasionally, and seemingly randomly, punctuated by a sudden trend toward less, less, less. I am thinking, of course, about the famous extinction episodes caused by meteors, ice ages, and events that we are still trying to understand. Extinction happens. Even without the help of modern technology, evolution and extinction are a pair. Evolution gives, and extinction takes away, but evolution rarely gets discouraged. It just keeps giving and giving until the wounds of extinction are healed and, indeed, we look around to see more species than ever before. The third thing we know about evolution is that it likes complexity. From life forms of a single cell, evolution has created organisms, like us, that have trillions of cells. From life forms that looked like little more than blobs, evolution has created things with feathers and eyelashes, fingernails and eyeballs, scales and fur. Although extinction happens, and usually takes out the most complex organisms first, the undeterred nature of evolution continues to develop and intensify the complexity of organisms—kind of like a Willy Wonka chocolate factory gone haywire. And here we have it, folks—the human. The most complex thing evolution has created so far. It walks, it talks, it reads, it writes. It can even understand . . . (drum roll) evolution itself!

This is the story of evolution from past to present. It contains direction and, to the human minds who have grasped it, a certain beauty and rightness. We like the direction, we find the story attractive, and although it may just be the human ego at work, in a way we do feel that evolution is all because of us, because the world had to make us, because we had to exist or the world would not be complete. This direction feels like God to us. The direction wanted us to live, wanted us to understand the why and how of evolution—and so we do. It's that miraculous Evolution Story, and we want to sing it and dance it and make paintings about it. We are the culmination. We are so special.

Although this is the attitude of many, a number of biologists object to viewing evolution in this way. Evolution has no direction, they will say; the notion of evolution progressing is just a human conceit. This appearance of progression is simply a result of chaotic change. The late Steven Jay Gould explains this argument in his book, *Full House: The Spread of Excellence from Plato to Darwin.* He argues that because life on earth began with extremely simple forms, organisms as close as possible to minimum complexity, we should expect a strongly skewed distribution of complexity with an essentially constant mode close to low complexity and a few high-complexity organisms. Over time, random drift alone will give just that sort of distribution—the type of distribution we see today. His argument is well made, but despite the force (chaos or God) behind increasing variety and complexity, in my mind there is indeed a progression.

Understanding all these things about evolution, what happens when we try to predict what evolution will create in the future? Suddenly there is a visceral fear. We get the feeling that things will start to go awry from this point on. Or have they already started going awry? The crops that grow in my neighborhood have all been genetically modified. The genes we have spliced into our specially patented seed corn have escaped into the wild ancestral corn populations of Mexico. This year, on the street where I live, a barn and a horse pasture have been replaced with a red light and a McDonald's. I liked my town better the way it was last year. And I liked the way it was ten years ago better than that, and I wasn't alive fifty years ago, but when I see photographs and hear stories of that time, I think I would have liked it even more back then.

I have not had the tests, but I'm sure that like 98 percent of American women, my body contains toxins such as dioxin, DDT, PCP, and various other endocrine disrupters. I'm also aware that I have probably passed these toxins to my daughter through breast milk. Fifty years ago, we didn't have so many toxins in our bodies. I know we have done well in cleaning up some of our messes, but habitat destruction, the number one cause of species extinction, is accelerating every year. Two hundred years ago, there were bears, wolves, parakeets, and gigantic trees on the land where I sit writing. Now there are none of those here. I would prefer if there were still native ecosystems intact, and I am not alone in these feelings. Most people I talk to feel that the places where they live are getting worse instead of better. So when we attempt to understand where evolution is taking us, where the future is taking us, we lose confidence in the process that we formerly viewed as good and right. It's a fascinating cocktail party question: "Do you think the world is getting better or worse?" Ask yourself first; then ask a few others.

In the late 1970s, there was a popular punk rock band called Devo, its name short for *de-evolution*. (Remember "Whip it Good"?) This band, from Akron, Ohio, was influenced by the writings of Oscar Kiss Maerth. Maerth believed that early hominids began cannibalizing each other, and by eating brain tissue, their sexuality and intelligence became artificially amplified. The human brain increased in mass, and consequently, over time, the human race suffered from an oversized sick brain that made man's self-destruction appear to be progress.[1] The only solution, the devos believed, was to demolish human culture entirely and begin again. De-evolution.

This view is in stark contrast to the worldview of the transhumanists, who advocate continuing and accelerating the transformation of the human condition through technological means—in other words, turning our tools on ourselves to create superhumans (posthumans). To be a transhumanist, one merely has to share their worldview. There is talk in this community of neural prostheses (computers implanted in the brain), "uploading" memories, and cybernetic immortality—the merging of human and machine.[2] The units of evolution are no longer biological genes, they argue; instead, they have become cultural information patterns. Evolutionarily speaking, they say, it makes no sense for our memories to die with us. We have spent a lifetime learning things that could be of value to other humans, present or future; therefore, we should strive to develop the technology to capture and transmit this information. The transhumanists see this shift as a natural part of the drive of evolution toward complexity.

In contrast to the transhumanists, who believe that humanity is improving (they cite increasing intelligence, longer life expectancy, and improved quality of life), the devos see us in the early stages of a human-induced extinction episode (they cite habitat destruction, persistent toxins, and nuclear weapons). It is a fascinating dilemma, really: Will our instincts toward power and immortality result in our salvation or our destruction?

The transhumanists dismiss those with a pessimistic outlook, claiming that it is a result of the stress on individuals and society caused by the accelerating speed of the change wrought by technological evolution. Of course, they see the changes in a positive light and counsel others to do the same. (We'll be colonizing other planets in no time!) The devos will have none of it, however, believing that the transhumanist worldview is shaped by a hominid brain gone bad, bent on enslavement and destruction of the earth. The devos claim that the transhumanists will finally be forced to realize that their existence is at risk by two phenomena: uncontrollable overpopulation and destruction of the environment.

The transhumanists claim that we can objectively measure progress by determining if our quality of life is improving or declining.[3] Did you know that there is a World Database of Happiness? It is compiled by sociologists who contribute articles to the *Journal of Happiness Studies*. They say humans are happier than ever before. Every variable used to measure quality of life—health, wealth, security, satisfaction, and others—shows improvement, and this, according to the transhumanists, means that we are making progress. The devos think it is our sickness that makes us want to continually "improve." We constantly set new goals for ourselves, which require us to work harder, but no matter how hard we try or how much we accomplish, we never find spiritual meaning or peace. We are the only animals, they say, who work in this self-imposed way. They consider this a symptom of our sickness. Maerth writes:

> The visible chain of his unsuccessful material remedies is what he calls progress, the invisible one is hidden in his soul: bitter disappointment and doubting hope. Criticism or doubt of his "progress" he always regards as sacrilege. He defends it as if he were under some evil spell, even if he has to suffer and work more under its burden. No load is too great for him if it is put on him bearing the hypnotically effective label of excellence, "Progress." And he himself places no limits to his progress. Does that mean that his troubles too are boundless? Or is this progress not intended to put an end to his troubles at all? . . . He has achieved nothing; on the contrary, he is blindly treading the path leading to his certain self-annihilation, and calls this path progress.[4]

If you want to be devo, you should fight this unhealthy urge to work and to progress. So the answer to the question: Is the world getting better? And consequently, one's own idea of progress depends on one's worldview. While back to Eden may be the most desired outcome for the devos, it would be hell on earth for the transhumanists.

The whole brain-eating-apes scenario of the devos is rather suspect, especially as it relies so heavily on the discredited concept of Lamarckian evolution (inheritance of acquired characteristics); but as a biologist, I also find major flaws with the transhumanist position. First of all, perhaps because most of the transhumanists are philosophers and not biologists, they discuss evolution only in terms of humans. Their notion of progress, measured by quality-of-life indicators, considers only human life. They do not seem to realize that we are supported by a great web of living organisms, that our very being depends on those other organisms, and that evolution, in whatever form, must happen in connection to other living things. I wonder what sort of quality of life we would see if we could query the fish in the sea, most of which are suffering from declining populations due to overharvesting and increased pollution; or the forest trees that are more and more frequently grown in monocultural plantations and harvested while young; or the monarch butterflies that return from their long migrations to find the ancestral meeting trees gone. We may soon be able to implant a neural prosthesis that contains the entire Library of Congress, but the human we implant it into will still need to breathe and eat. And we depend on other organisms to provide the materials for both of those functions and many more. The introductory biology textbook I teach from lists seven major theories of biology. The newest is the theory of ecosystems: "Organisms interact with each other and with their environments. Changes in any part of the biological community or the physical environment invariably cause changes in other parts."[5] When the transhumanists finally get around to admitting that there are environmental concerns, they tend to quote cornucopians, such as the late Julian Simon or Bjorn Lomborg, whose rosy predictions have been discredited by most well-known ecologists. One gets the impression that the transhumanists are all male, all white, and spend all day with their computers. Hey guys, try going camping this weekend and leaving the laptop home. There's a big world out there.

Another consideration for those advocating transhumanist thinking is the Fermi Paradox.[6] It goes like this: If there are many suns with planets, as we now know there are, and if a number of these planets are Earth-like and have life (unknown), and if life eventually evolves toward technologically advanced beings that can travel through space (unknown), why don't we have proof that space travelers exist? Two possibilities are that (1) we will be the first beings to evolve to that level of sophistication, or (2) maybe every civilization that develops to a certain technological level causes its own extinction.

Ah, yes, the question of *should* given the *could.* In an article that analyzes possible human extinction scenarios, Nick Bostrom, cofounder of the World Transhumanist Association and professor of philosophy at Yale University, points out that the most serious risks of extinction are generated by the technology of an advanced civilization.[7] Hmmm—. By the way, his current pick for "most likely to cause extinction" is misuse of nanotechnology.

Then there are the claims of the devos to address. What about that uncontrollable human population growth? Will every one of our (soon) 10 billion get to be superhumans? If not, what happens to the others? Personally I am concerned about supplying food and water to those folks; forget the silicon chips. Oh, that's right; you say that with our "implants," we will be able to figure out a way to make more food and water. Maybe. Maybe not.

So if I'm sounding like such a know-it-all, what do I think the evolutionary future will hold? First, we must admit that evolution is no longer operating independently. For the first time ever, one of evolution's creatures has partial control. As soon as humans started "breeding" other organisms, we started shaping evolution to our own ends. This work has just become more sophisticated with gene splicing and designer crops, and our unnatural creations are taking over more of the Earth's surface every day. Even our own species no longer evolves through survival of the fittest; now it is survival of the richest. Mother Nature won't let you reproduce? No problem; just put $50,000 down and sign on the dotted line. Just when animals with brains big enough to comprehend the genetic code have evolved, the very same big-brained creatures have short-circuited the process that created them. We are scrapping a process with billions of years worth of checks and balances for a great experiment. There are so many unknowns in what we are doing now that no geneticist claims to know how it will turn out.

Don't sound so gloomy, you might be thinking; *look at the crops we can grow, the milk we can produce, the diseases we can cure, the organs we can transplant. This is good, right?* And the answer is yes; it is very good for humans—for now. But we are enslaving the planet for our own pleasure. The species we currently eat, build with, or keep as pets are given lots of space and attention; but many of the others, such as wild orchids, tigers, and bats, are driven closer to extinction every day. There will be no more evolving for them.

I think the transhumanists are right about one thing: we will create superhumans. We will use technology to extend our biological capabilities. We cannot help ourselves. In fact, it has started already with cardiac pacemakers and laser eye surgery. It may be wonderful for those humans who can afford the microchip brain implant, but it's going to suck for all other living creatures. The "enhanced ones" are likely to bring a whole lot down with them when they go. Bostrom calls this scenario a "whimper": "A posthuman civilization arises but evolves in a direction that leads gradually but irrevocably to either the complete disappearance of things we value or to a state where these things are realized to a miniscule degree of what could have been achieved."[8] There will be whimpering all right.

Evolution has been progressing in one direction for billions of years. What arrogance to assume that we could influence it, we reassure ourselves. Whatever is happening now must be part of the plan, it must be evolution in action, it must be nature moving toward more species and more complexity. Right? We are part of nature, aren't we?

Good question.

Notes

1. Oscar Kiss Maerth, *The Beginning Was the End* (London: Joseph, 1973).
2. Francis Heylighen, "The Future of Humanity," in F. Heylighen, C. Joslyn, and V. Turchin (eds.), Brussels: Principa Cybernetica, 1999. <http://pespmcl.vub.ac.be/Futevol.html>.
3. Francis Heylighen and Jan Bernheim, "Global Progress I: Empirical Evidence for Increasing Quality of Life," *Journal of Happiness Studies* 1:3 (2000). <http://pespmcl.vub.ac.be/Papers/Progress.html>.
4. Maerth, *The Beginning Was the End,* 100.
5. Nancy Pruitt, Larry Underwood, and William Surver, *Bioinquiry: Making Connections in Biology* (New York: Wiley, 2003), 11.
6. G. D. Brin, "The 'Great Silence': The Controversy Concerning Extraterrestrial Intelligent Life," *Quarterly Journal of the Royal Astronomical Society* 24 (1983): 283–309.
7. Nick Bostrom, "Existential Risks: Analyzing Human Extinction Scenarios and Related Hazards," *Journal of Evolution and Technology* 9 (2002). <http://www.jetpress.org/vol9/risks.html>.
8. Ibid.

Moral Progress

Dale Jamieson

It is presumptuous for a philosopher to write an autobiography. What a philosopher has to offer is coherent claims and cogent arguments, and these have little to do with personal circumstances. For claims and arguments—except perhaps those involving words such as *he, she, here,* and *now*—are neutral with respect to person, place, and time.

Sometimes this is what I believe. But despite this, I often find myself turning first to the "notes on contributors" sections of books and journals. When I do this, a certain ambivalence often bubbles to the surface. On the one hand, I often feel frustrated when I can find nothing about Jane Doe's background, education, and professional position. How I read her article on theories of justice depends in part on where she studied, with whom, and whether she has a degree in law or economics. On the other hand, I find it in bad taste, sometimes even infuriating, to read that Doe is a triathlete whose hobbies include growing orchids and racing formula one cars. Who cares?

Still, some of the most interesting and important autobiographies have been written by a diverse group of philosophers that includes Augustine, Rousseau, and Russell. Not only are these autobiographies good literature; they are also important to the interpretation and defense of their authors' philosophical positions. The autobiography of John Stuart Mill, the greatest of utilitarian philosophers, poignantly makes the case against the livability of his own philosophy, at least when reduced to its crudest form. The twentieth-century Harvard philosopher W. V. O. Quine's recital of the history of his bodily movements gives us a glimmer of what it is like to live as a behaviorist.

Somewhere in these ruminations may lurk a defense of philosophers' autobiographies. However, such a defense would do little to justify an excursus into the story of my life. David Hume's autobiography is important because he was the greatest of eighteenth-century philosophers. I cannot even claim with confidence to be the most important philosopher to have emerged from my working-class neighborhood in San Diego, California.

There is a different consideration that propels me toward autobiography. I cannot help but feel that there is something dishonest about the abstractness of most philosophical writing. Anyone who dares to tell people what they ought to do or think, whether Immanuel Kant, George W. Bush, or me, should be willing to make public his or her name, address, and telephone number. Telling the story of one's life is one way of doing this. Another reason for these reflections is that much of my work in one way or another concerns moral progress. The story of my life may help to illuminate these concerns.

I was born in Sioux City, Iowa, in 1947. My father worked in a bakery, but many of the neighborhood dads worked in the packinghouses (as they were euphemistically called). Terrible smells often wafted over our neighborhood, and it seemed odd to me that the men who worked in the packinghouses did not talk about their jobs. As a child, I did not associate the stench with blood, nor did it occur to me that these men killed animals for a living.

Every summer we visited relatives in California, and, finally, when I was about twelve, we moved to San Diego. We were part of the massive post–World War II migrations that "rationalized" the American labor market.

When I was fourteen, I was enrolled in the Academy of California Concordia College, a Lutheran boarding school in Oakland, California. Both of my parents had been forced to drop out of high school due to various circumstances, and while our family had great respect for education, we had no tradition of learning for its own sake. During those early teenage years, I conceived strong religious convictions, and my education, which involved enormous sacrifice on the part of my parents, was justified on the grounds that I was preparing for the ministry.

While at Concordia, I absorbed two major influences. The first was the moralism of the Lutheran tradition, which borders on self-righteousness (and in my case often transgressed that border). The second was the generally antiestablishment, countercultural attitudes characteristic of the San Francisco Bay Area in the early

1960s. When we weren't studying Greek, Latin, and (of course) German, we were hanging out in the juke joints on East Fourteenth Street in Oakland or in the bookstores on Telegraph Avenue in Berkeley. Seeing the police drag Mario Savio out of Sproul Hall at Berkeley during the Free Speech Movement and watching young black men morph into the Black Panther Party for Self-Defense was part of daily life for an adventurous Oakland teenager in the early 1960s. It should not be surprising that I became an active participant in the civil rights and anti–Vietnam War movements.

When I was kicked, pushed, and manhandled outside a Lutheran church while distributing antiwar leaflets, it became apparent to me that my future was not in the Lutheran ministry. I drifted through several Bay Area colleges before winding up at San Francisco State, where I thought that I might study filmmaking or creative writing. However, there was a matter of some embarrassment that needed to be cleared up: although I was certain that the Vietnam War was immoral, I wasn't sure exactly what that meant. I thought that if I took an ethics class with Anita Silvers, then I would find out. I didn't. Instead I got hooked on philosophy.

In the fall of 1968, Ronald Reagan and S. I. Hayakawa declared war on students and faculty at San Francisco State for the crime of demanding an ethnic studies program (admittedly in a rather rude and impolite way). Nearly one thousand students and faculty were arrested in a single day for simply being on campus. By this time, I had seen enough beatings and gassings, and I had discovered Wittgenstein. Through the largesse of a third cousin who had died childless, I went to study at Birkbeck College in London. The next year, I returned to San Francisco to take my degree and figure out what to do with my life.

Fortunately, I didn't have much time to think about it. My teachers in both England and San Francisco agreed that I should go to graduate school. My teachers in London were impressed not because I was brilliant at philosophy, but because this long-haired San Francisco hippy could do it at all. My teachers in San Francisco thought well of me because I had studied successfully in England, which was then still the center of "analytic" philosophy. The idea of graduate school sounded pretty good since the only other option I had considered was borrowing money to start a car wash or a laundromat. (I thought that these were businesses in which you showed up for work only to take money out of the machines.) My brilliant career was soon underway.

I had the following criteria for choosing a graduate school: I would go wherever they offered me the most money, so long as it wasn't in California or in an urban area. For me, going to graduate school was going back to the land with a safety net. All signs pointed to the University of North Carolina at Chapel Hill, where I could study with W. D. Falk, who was a friend of Ruby Meagher (my London tutor), and Paul Ziff, who was rumored to be someone whose brilliance was exceeded only by his eccentricity. I could also live in the woods.

It worked out. I bought some land and some tools, and built a simple house in a hardwood forest. I also did some philosophy.

Shortly after arriving in Chapel Hill, I discovered that graduate students who wanted to be taken seriously did not work on ethics and aesthetics. These subjects were regarded as appropriate only for people who were incapable of doing serious analytic philosophy (generally females). Not wanting to be thought one of them, I worked mainly in philosophy of language and linguistics and in philosophy of science and mind.

In 1975, I was writing my dissertation when a friend who was teaching at North Carolina State suddenly quit when his band got its big New York break. I soon had my first academic job; since then, I have been continuously employed as a college professor, teaching at nine colleges and universities on three continents.

For years, I prided myself on being a rootless cosmopolitan, at home wherever there were good books and music, cheap but decent wine, and the companionship of similarly disaffected bohemians, preferably of working-class origins. Some of this began to change with age and responsibility, but the effects of fatigue and friction cannot be ignored. Having seen myself for years as the man who fell from space, I began, alarmingly at first, to recognize myself in my parents.

And so in the early 1990s I found myself at a family reunion near the homestead that my great-grandfather had farmed. In 1872, August ("Augie") Kienast immigrated to the United States from a village near what was then Konigsberg in East Prussia and is now Kaliningrad in Russia. After working as a lumberjack in Michigan and traveling throughout the West, Augie settled in an isolated immigrant farming community near Atlantic, Iowa. This community was defined by its pietistic Protestantism and was intensely loyal to German culture. German remained the language of the community until the 1950s, even for public education, despite periodic attempts to suppress it.

I attended the reunion with my mother, the daughter of an Irish Catholic and the granddaughter of Margaret Ryan, about whom it was said could not utter the word *Protestant* without prefixing the phrase *goddamned,* which she meant literally. Many in Augie's community probably would have returned the compliment. By

1947, these sentiments had subsided sufficiently so that when my parents married, although their religious differences mattered (my mother converted to Lutheranism), they did not engender serious conflict. Now, surrounded by several generations of Augie's progeny, I could not help but reflect on how much more accepting of religious differences Americans had become. Intermarriage between Catholics and Protestants, which would have been intolerable to many of Augie's contemporaries, barely raises an eyebrow, much less an objection, among his descendants. As I watched several generations of Kienasts playing with the newest member of the family, Bradley, an African American child adopted by one of my cousins, I wondered whether the contemporary American obsession with racial differences would come to be viewed as dangerously atavistic as a fixation on religious differences appears to us now. From the perspective of the Kienast family reunion, moral progress seems palpable.

From another perspective, however, it is easy to be skeptical. In this century, more people have died in wars, famines, and preventable disasters than in the rest of history combined. For the first time, we have developed technology that allows us to threaten the very existence of our own species as well as much of the rest of the biosphere. Even today, religious differences are implicated in mass killings and atrocities, as the attack on the World Trade Center made vivid to many of us. While there is a rising tide of concern about human rights, the treatment of animals, and the state of the environment, this may simply reflect the hope for something better rather than any real progress.

And yet I cannot help but believe in moral progress. I began my professional career as a philosopher of language but returned to moral philosophy in part because I wanted to make a difference in the world. The catalyst was meeting Tom Regan when I first took up my job at North Carolina State. Tom had just completed an essay entitled "The Moral Basis of Vegetarianism," which had been accepted by the *Canadian Journal of Philosophy*.[1] Having grown up on friendly terms with various dogs, fish, birds, and rodents, I was less invested in animal exploitation than many other people. My best friend was a dog named Ludwig, and I had been vegetarian since 1972, mainly for health reasons, but also to avoid offending my (then) girlfriend. Still, Tom's paper blew me away. It seemed wild to suppose that one could give a rigorous, analytical argument implying that much of what people do is grossly immoral. Yet after some weeks of resistance and struggle, I admitted that Tom was largely right. We became comrades, and my work began to move in the direction of moral philosophy. How cruel it would be to discover, a quarter century later, that moral progress is impossible. I might as well have continued the battle against possible world semantics as devoting myself to protecting the interests of animals and nature.

Whatever the truth about moral progress, there is no denying that the language of moral progress has become ubiquitous in Western societies, whether as a cover for invading small countries or in the service of genuinely progressive ideals. However, Western societies have not always been centrally concerned with this idea. There is scholarly debate about whether the ancients and medievals had the idea of progress (as opposed to the idea of degeneration) and, if so, how influential it was. There is, however, a high degree of consensus that this notion moved to center stage with the work of such thinkers as Descartes, Fontenelle, Perrault, and St. Pierre. Scholars typically date the concern with progress to the late seventeenth or early eighteenth century. In *History of the Idea of Progress,* the sociologist Robert Nisbet writes that between 1750 and 1900, the idea of progress was "the dominant idea [in the West], even when one takes into account the rising importance of other ideas such as equality, social justice, and popular sovereignty."[2] The historian Crane Brinton writes in *A History of Western Morals* that in the eighteenth and nineteenth centuries, "most Westerners were confident that human beings everywhere were getting morally better as well as materially better off."[3] The Reign of Terror in the last stages of the French Revolution may have led to some doubts, but these events were confined to a ten-month period and claimed only about 20,000 lives in a population of 22 million. Brinton acknowledges that there was persistent underground opposition to the idea of progress from disgruntled intellectuals (e.g., Ruskin, Arnold, and Dostoyevski) who seemed to think that material progress necessarily implied moral decline. Still, it is reasonable to think of the period from the signing of the Declaration of the Rights of Man in 1789 to the outbreak of World War I in 1914 as the great age of belief in moral progress. This came to an end on Flanders' fields.

Still, the question of moral progress remains deep in the European psyche. At the dawn of a new century, the horrors that Europeans inflicted on each other during the first half of the twentieth century seem like ancient history. Yet the intense European reaction to the American callousness about inflicting death, as both a form of punishment and an instrument of foreign policy, suggests the anxiety of suppressed recognition. This sense of guilty familiarity may also help to explain European compassion for the developing world compared to American indifference. The brutalities of colonial history were so extreme that in some cases, they amounted to what has been described as a holocaust. Without fully acknowledging their responsibility, many Europeans now express a sincere willingness to aid developing countries. Americans, without a colonial legacy of the same extent and explicitness, are more likely to view the developing world as irrelevant, except when terrorists emerge to "attack freedom." This attitude is dangerous for many reasons, not least because it feeds a tendency to say "a plague on both your houses" when confronted by wars and atrocities in the developing world.

The case for the European Union is to a great extent a moral case. By taming the historical rivalry between France and Germany, it is seen by many as a way of preventing another world war. Indeed, Europe is now the center of thought and action for advocating a vision of the world in which universal human rights prevail, and nations are governed by international norms that keep the peace, guarantee environmental protection, and prevent global hunger.

The idea of moral progress is also deeply embedded in American culture. American exceptionalism is founded on the idea that America is "the city on the hill" whose mission is to bring light to the other nations of the world. Occasionally American leaders admit to well-intentioned "mistakes" or "failures," but the language is almost always one of self-correction and constant improvement. I can't think of an American president in my lifetime who was reelected (thus excluding Carter) for whom the mantra "our best days are yet to come" did not figure centrally in his political rhetoric. While America may appear to the world as a bastion of materialism and greed, the sense of optimism that these political leaders convey is not just confidence in the economic future of the nation, but also faith in its moral mission.

In the light of this history, it may seem surprising how much disagreement there is among thoughtful people about the prospects for moral progress. Reflecting on his fifty years in journalism, retiring *New York Times* columnist Anthony Lewis recently said, "I have lost my faith in the idea of progress . . . in the sense . . . that mankind is getting wiser and better. . . . How can you think that after Rwanda and Bosnia and a dozen other places where these horrors have occurred?"[4] Yet the contemporary philosopher and cultural critic Richard Rorty seems to take moral progress for granted, suggesting that its pace has increased over the past two centuries: "The nineteenth and twentieth centuries saw, among Europeans and Americans, an extraordinary increase in wealth, literacy, and leisure. This increase made possible an unprecedented acceleration of the rate of moral progress."[5]

The historical record is undeniably equivocal and tragically ironic. The French Revolution produced both the Declaration of the Rights of Man and the Terror. World War I, the "war to end all wars," led to the bloodiest war in human history.

In their attempts to stop one of the most inhumane regimes in history, the English-speaking democracies unleashed a total war against civilians and ushered in the nuclear age.

Piling up such anecdotes proves little. Still, it is hard to read the historical record and not form some impressions. The rhetoric that promotes peace, human rights, and respect for nature appears to be better entrenched today than previously. Rather than being the language of reformers and radicals on the margins of society, these sentiments are now voiced by some of the most powerful people in the world. The cynic will say that what this shows is not the progress of morality but the growth of hypocrisy. The same old corrupt behavior is described in prettier terms. Still, there is a case to be made for the idea that language matters. What people do is important, but it also matters how they justify their actions and what ends they claim to seek. The cynic will point out that while, on the whole, softer language may have become more prevalent, this kinder, gentler vocabulary has been available for a very long time. One has only to be reminded that the Golden Rule in its various formulations occurs in many ancient traditions to see that the language of moral progress was in place long before people propagated their greatest horrors.

Still, moral and political change in the past several centuries does seem to be moving away from hierarchy toward greater egalitarianism. The differences between people that were enforced in Europe and America well into the twentieth century would be intolerable today. The idea that a government could exclude most of its citizens from voting and still be representative sounds like a joke. It also appears that there has been a softening in human cruelty. Until the eighteenth century, the atrocities that occur today in ethnic and tribal wars were the norm rather than the exception. The cruel forms of public execution and the merriment that they evoked occur nowhere in the world today. Even the common forms of amusement in early modern Europe–bear baiting, cockfighting, and so on—continue only in attenuated and atavistic forms. Perhaps we have become better at face-to-face encounters. Yet technology has given us the power to engage in remote destruction on a scale that is unprecedented. The amount of killing of humans, animals, and the rest of the biosphere that we have engaged in over the past century would have been unthinkable by even the greatest of ancient tyrants. And although our kind of killing is often represented as precise, targeted, even "antiseptic," anyone who has really tried to understand what goes on in modern slaughterhouses or contemporary warfare will come face to face with the same old horrors: people and animals burned alive, flayed, left to die in abject misery with little or no comfort. I am inclined to agree with Brinton when he writes, "We may indeed be no worse morally than our ancestors, and we may even be on the average a bit less cruel, less brutal, but grouped in nation-states we can do a lot more evil more quickly and more efficiently than they could."[6]

Still, the Enlightenment ideology of *fraternité, egalité,* and *liberté* remains deep in our bones and rhetoric. When we look back at institutions of slavery and sexual subordination that were the norm only a heartbeat ago, it is hard not to feel smug about the progress that has occurred. But, on the other hand, the horrors of the twentieth century seem to impel a kind of postmodern cynicism. Language changes but life goes on, and life in society is fundamentally about power, difference, and oppression. Americans are horrified by the Chinese eating dogs, while the Chinese don't understand how Americans can eat pigs. The Romans were horrified by the human sacrifice practiced by the Celtic tribes in Gaul, but the carnage of the Roman games did not seem to cause a flicker of conscience. Or so it is said.

These arguments are worth having because they remind us of how much is in play when we consider the question of moral progress and how different the question looks depending on our focus. Questions about moral progress are fundamentally pragmatic. They do not come out of a vacuum. Recognizing this may put to rest the nagging suspicion that those who make such claims on behalf of themselves or their societies are arrogant in suggesting that they are morally better than past generations. What such people claim is that in their own light, they are better on some dimension than those who have come before. This modest view of moral progress may disappoint those who lust for a grand narrative that would sweep through time, space, and society, leaving a clarified vision of humanity in its wake. But of course, such grand narratives are themselves often the source of staggering amounts of human misery. If what I have said is correct, then we must think locally and contextually about moral progress rather than universally and unconditionally.

In my lifetime in the United States, there are two examples that seem to provide clear cases of moral progress. One is the struggle against American apartheid, which came to a head in the 1960s; the other is the animal rights movement, reinvigorated in the 1970s. In different ways and with various degrees of focus and clarity, both of these movements are still in motion.

About the time that I was born, Harry Truman, an unelected president from a border state that was traditionally aligned with the old Confederacy, took a major political risk. Under pressure from civil rights leader A. Philip Randolph, he issued an executive order outlawing racial segregation in the nation's military. It is difficult today to recapture the importance and courageousness of Truman's action. Although they were forced to serve in segregated units under the command of white officers, African Americans had fought bravely in World War II. It might seem that such a demonstration of loyalty and courage would have been enough to convince almost anyone that blacks should be granted equality, at least in this limited domain. For that matter, it might seem that integrating the military was a very small step toward justice in a society that was still lynching young black men. Yet this decision almost cost Truman the 1948 election and had far-reaching consequences. Six years after Truman's order, the U.S. Supreme Court declared that segregated schools were inherently unequal, thus effectively requiring school integration throughout the nation. In the wake of this decision, a new generation of black leaders came to prominence, leading demonstrations, boycotts, and campaigns of civil disobedience in the fight for racial justice. In 1964, Lyndon B. Johnson, another unelected president, this one from one of the states of the old Confederacy, persuaded Congress to pass the most sweeping civil rights law since Reconstruction. This was followed by a battery of other laws, and in 1967 by the appointment of Thurgood Marshall to the Supreme Court. Marshall, the first African American to serve on the Court, had argued the school desegregation case before the Court thirteen years earlier. By 1970, the legal structure of apartheid had been abolished, affirmative action programs were being instituted, and the national consciousness regarding race had been radically altered. Although racism had not been abolished, it was now shunned and outlawed in its most overt and blatant forms.

Consider another story. Several years after the first Earth Day in 1970, books such as Peter Singer's *Animal Liberation* and articles like Tom Regan's "A Moral Basis of Vegetarianism" began to appear.[7] Many people who thought of themselves as politically progressive began to see the abuse of animals and nature as part of the same structure of oppression that produces racism, sexism, and class domination. Some, inspired by the nonviolence of the civil rights and antiwar movements, saw vegetarianism as a natural extension. What had been sleepy animal welfare organizations, staffed mainly by people with sentimental attachments to animals, almost overnight became animal rights groups, sometimes militantly directed toward changing the moral and legal status of animals. This movement has gained some victories, but on the whole it remains very much a work in progress.

What has changed is the consciousness about animals, at least in many sectors of society. Many people now find themselves sharing meals with moral vegetarians, perhaps for the first time having to acknowledge in some way that using animals for food stands in need of justification. To some extent, what happens in laboratories and slaughterhouses has been brought to light. Those who retain "Old McDonald" fantasies of how farm animals are treated now border on denial or culpable ignorance. Real behavior change has not kept up with this growth in awareness, and the contradictions are sometimes painfully obvious. But the recognition of contradiction is itself a sign of moral progress, at least compared to the moral complacency that governed our treatment of animals prior to the 1970s. While widespread legal reforms have yet to occur in the United States, the European Union is in the process of transforming the conditions under which hundreds of million farm animals live.

In light of these stories, it seems plausible to suppose that moral progress with respect to a subordinated group has four stages. The first stage involves recognizing the practices of subordination as presenting a moral issue as opposed to presenting questions of taste, etiquette, or personal preference. Having gotten this far, we might take a paternalistic interest in defending those being badly treated; we come to see them as objects of morally admirable charity. The next stage is to recognize that rather than being only the objects of charity, those who have been subordinated have rights not to be harmed. Finally, we may come to see them not only as bearers of "negative rights," but also as bearers of "positive rights," entitled to what they need in order to flourish.

How these changes occur is an important but neglected area of inquiry, and what I have to say here will be quite tentative. As a beginning, it is helpful to distinguish between appeals that are launched from within conceptual frameworks and those that originate from without.

Martin Luther King, Jr., often appealed to principles taken to be foundational in America, such as "all men are created equal." When protesters were arrested for sitting in at "whites-only" lunch counters, he argued that it was they, not the racist sheriffs who arrested and beat them, who were acting in the spirit of America. The problem was not that some young black people wanted to be served at a lunch counter; it was that American law and morality did not prevail in the states of the Old South where service was being denied. A principle such as "all men are created equal" gains its power from the recognition that everyone has a stock of fundamental interests, and to elevate the interests of some at the expense of others, simply on the basis of skin color, is indefensible. However, progress leads to further challenges, as the very language of the principle, "all *men* are created equal" suggests. What is objectionable is not only discounting the interests of some on the basis of race but generally discounting interests on the basis of any morally arbitrary property, including gender. But isn't species membership also such a morally arbitrary property? This question suggests others. Why are millions spent to keep alive a severely brain-damaged human whose interests are not as urgent or expansive as those of the dog who is being used for practice surgery? Why, for that matter, are some dogs honorary humans, while others are disposable entertainment or food?

External assaults on a conceptual framework that supports subordination often occur when internal appeals have failed. For example, frustrated by the failure to persuade the supporters of slavery by rational argument or appeals to human decency, nineteenth-century abolitionists often aimed at something like religious conversion, sometimes seen as the only hope for changing people who had been corrupted by slavery. Similarly, animal rights activists often try to reorient people's view of animals—to see them as complex, intelligent creatures rather than as defective humans, governed only by instinct. Environmentalists try to bring people to see swamps as wetlands and "nature red in tooth and claw" as Darwin's "entangled bank." This can be risky business. Those who pursue this strategy often sound "crazy"—until (and unless) they succeed.

These strategies are often employed in tandem. Often, inconsistencies become clearer and arguments become stronger if questions are reframed and reconceptualized. At the same time, there is no reason to think that every case of moral change proceeds in the same way, or that I have even begun to touch on the range of strategies that are available.

For those of us who want to improve the world, examining the civil rights and animal liberation movements is important because they provide models of moral progress and inspire us to feel part of a struggle directed toward world historical change. From this position, we see ourselves as actors in a universal drama. The changes for which we struggle will take centuries to fully effect and consolidate, and we will be gone before our most profound goals are realized. This realization does not make our day-to-day frustrations—much less the suffering of a single human or animal—any more acceptable. However, such a long-term perspective may help us to see that although eliminating human and animal suffering is an urgent and demanding obligation, it is one that we are more likely to discharge successfully in a spirit of compassion, understanding, and humility than with dogmatism and vindictiveness.

Ultimately, we live short lives compared to human history, and in small neighborhoods compared to the global community. It is from this point of view that our motivation is gathered. Small moments of success must be savored and celebrated. Thus, I return to where I began and to the Kienast family reunion. The moral progress that is palpable in the Kienast family over several generations is real and important. This, as much as anything else, makes me feel that it is worth living one's life in pursuit of moral progress, even in the tortured, precious way of a philosopher.

Notes

1. Tom Regan, "The Moral Basis of Vegetarianism," *Canadian Journal of Philosophy* 5, no. 2 (1975): 181–214.
2. Robert Nisbet, *History of the Idea of Progress* (New York: Basic Books, 1980), 171.
3. Crane Brinton, *A History of Western Morals* (St. Paul: Paragon House, 1990), 414.
4. Anthony Lewis, quoted in Ethan Bronner's, "After 50 Years of Covering War, Looking for Peace and Honoring Law," *The New York Times* (December 16, 2001): Weekend 9.
5. Richard Rorty, "Human Rights, Rationality, and Sentimentality," *Truth and Progress: Philosophical Papers* 3, (New York: Cambridge University Press, 1998), 167–185.
6. Brinton, *History of Western Morals,* p. 443.
7. Peter Singer, *Animal Liberation* (New York: New York Review of Books, 1990).

Coevolutionary Flashes in the Withering Beam of Progress

Richard B. Norgaard

It is an early Saturday in December, and I am in Uppsala for a few days rather than somewhere else. The midday sun is low on the horizon as I head for the oldest cathedral in Scandinavia. Walking past buildings stuccoed in traditional deep reds, greens, and yellows with their south walls basking in the tingling sunlight, the castle on the hill against a crisp blue sky, the bridge coated in ice condensed from the stream's mist. Upon entering the cathedral, I find once again that Catholicism has its local particulars. The vastness is light, the pews birch blond, the walls and marble caskets as clean-lined as modern Scandinavian furniture. Out into the cold again, I stroll past the university, the academic home of Celsius and Linnaeus, and then head into the city center to mingle with early Christmas shoppers. Main Street, reclaimed for foot traffic, is lined with tasteful shops selling domestically produced clothes for the rich. The department stores sell linens made in Portugal, shirts made in China, and shoes made in Indonesia, just like everywhere else in the once industrialized North. I step determinedly past a McDonald's overflowing with happy shoppers, reassuring myself that local bakeries, cafés, and delis are doing a brisk business today too. Emerging from a gallery of intermingled smaller shops, I hear the familiar music from the Andes and seek out the four men from Peru.

They are well positioned near an intersection, strumming, drumming, wailing on flutes, and singing their hearts out. Like me, they are in Uppsala rather than Barcelona, Berkeley, Chicago, Frankfurt, New York, São Paulo, Sydney, Tokyo, Toronto, or the other places I have heard their *companeiros* in the past few years. Sharing a millennium of culture, crassly making a buck, joining a multicultural global discourse, freezing their butts off, sending money home to Peru to sustain the indigenous resistance to the cultural imperialism of the very people gathering supportively around them—all of the above. And now I am also more aware that the bundled people strolling past me include a surprising number of colorful faces from around the globe, political and economic refugees from failed efforts to prompt progress, evidence of Swedish willingness to provide a Band-Aid™ to the badly bungled world.

I am in Uppsala today, but walk the streets of any significant city and you find the local, regional, and global sensuously intermingling. Different peoples are pulled by opportunity, pushed by failure, yet they are together, maintaining much of their identity. Social tensions are palpable as we face new cultural configurations. We are going in different directions: rich and poor; along diverse religious traditions; some clinging to the promises of scientific rationality, some returning to fundamental guidance, some seeking new paths to collective understanding. There are more than sufficient tensions for a plethora of local meltdowns. And yet we hold together while on multiple courses.

In the streets watching people, I find hope, even joy, amid our diversity, joining in an intricate dance of everyday life. Beyond the flow of people, but still in my face, however, I am struck by the ugly unfolding of petropolitics: George W. Bush, multinational oil companies, Middle Eastern dictators, and the followers of Wahhabiya. The cracks in nuclear power plants that were belatedly reported by trusted corporations to Japanese authorities complement the potential of decentralized dissidents to perpetrate biological warfare. I am aware of economic immigrants drowning in the Mediterranean and parching in the Sonoran Desert, as well as the greedy creativity of market entrepreneurs and their "public" accountants. Adam Smith's simple idea about the gains of trade has taken over the world, while the much richer understandings of climate scientists are denied by key leaders. This global picture of history gone chaotic is as terrifying as the diversity on our local streets is hopeful.

The future was not supposed to turn out like this. Progress was supposed to bring us to some cohesive, higher collective form. We were supposed to merge smoothly into one culture organized around a unified scientific worldview. As the new millennium begins to unfold, we see neither organized cohesion nor scientific rationality.

Certainly by now, the European identity should have dominated. Scandinavians should not, on the one hand, be so parochial and, on the other, have so clear an international voice. Other Europeans should not be reidentifying as Basques, Catalans, Padanians, and Silesians while interacting globally. The cultural contradictions are just as sharp in India, North America, and Southeast Asia. At every locale, who we are has frenetically unfolded into a multidimensional maze of locally fragmented, globally interconnected, mixed geographies and histories.

Science has pushed on by studying smaller and smaller parts, by branching and branching again. Some scientists are reconnecting some of the pieces, intertwining a young branch here with a pliable sprig there, but the connectors, working at making numerous different linkages, also cluster into particular groups and speak distinct tongues. There is no whole. Individual scientists know more and more about less and less. They have the time neither to read about what is happening beyond their area of expertise nor to explain what they are learning to others. While science as a whole is a fractious morass, corporations have been able to organize the parts of science they need to develop new technologies and bolster their profits. New technologies impose new risks and environmental costs, but the public funding of science, especially funding to build the cohesive understanding needed to manage the public consequences of private profiteering, has not kept up. Without such broader public comprehension, corporations are free to deploy fragments of knowledge, transforming the social and the natural while invoking the name of progress.

Across this chaos, however, some patterns are very clear. The rich are getting richer faster, consuming more and more, and the poor are not catching up. Indeed, an unfortunate number of people during the last quarter of the twentieth century dropped behind in absolute terms. Taking population growth into consideration, the desperately poor number about the same as when we set out to develop all nations fifty years ago. Progress was supposed to erase these material disparities, not provide rose-colored glasses for the rich. Political power is again proportionate to material control, after a modest tempering by democracy and science. Meanwhile, carbon sequestered over millennia as coal, petroleum, and natural gas is being burned to fuel our economies, recombining with the oxygen it once released. Increasing atmospheric carbon dioxide is like wrapping the globe in a blanket, more heat is being radiated back to earth rather than out to space, and so the globe warms. The diversity of life—already being squandered by the advance of monocropping, an acceleration of species mixing across previously more isolated ecosystems, and the introduction of toxins—is threatened further by climate change. And yet another trend is flat. The happiness of the rich is not going through the roof as the top 5 to 10 percent of the world's population consumes ever more. In fact, it is hard to find evidence that their happiness has increased at all.

These things are as clear as the denial is great.

And where are we going from here? The dominant trends cannot continue. To stop our negative inclinations, we will have to work together. Yet we are going in multiple directions. Common language, common cosmology, and common criteria are going to be necessary. How can we even understand what *better* might mean pursuing different trajectories? Historically, the vast light of progress, with its emphasis on scientific advance and a common human destiny, beamed ahead. Can we still draw on this waning torch to guide us adequately to the future? Or could a new way of seeing the human saga emerge to make sense of the cacophony and give life meaning, dignity, tolerance, and a path toward living with nature again?

Our histories were stories of human progress. Advancing scientific knowledge drove new technologies and the production of ever more and more exciting material goods. History unfolded around the steam engine, electricity, and then information technologies. Our future was in the hands of scientists and engineers. The Great Depression brought economists to the fore. With the successful reconstruction of Europe following World War II, the idea of material development on a global scale arose. Thereafter, progress needed economists to determine how nations should be configured to incite capital flow and clear the path for increasing global trade. Now we see genetic engineering calling us into a wildly new material future with exciting opportunities for scientific and entrepreneurial ingenuity. But the call is not so enticing anymore. We see the corporate lips on the trumpet. Our modern past has had too many social problems, even some atrocities, too many environmental missteps, some disasters. So as progress calls, we whisper other stories to each other. Materialism through science, technology, and economics did not simply make life better. Indeed, we can tell so many different stories about the past that the future is very confusing. And yet now, ironically, the old idea of progress, however dimmed, is about all we share—all that binds us together when it comes to talking collectively about the future.

How did we get here? Our old sense of progress has many deep roots. Of these, certainly the Christian belief that we should, could, and would improve our moral lot was a central one. As the prospects for material advance arose during the Renaissance, Christianity accommodated a cultural sea change, a gradual one punctuated by storms. The argument that greater material well-being would both reduce the temptation to steal and provide more opportunity to study the Bible fostered a separation between the moral and the material while positioning them as complementary partners. Thus, personal and collective moral advance still defined the end of progress, but material improvement was incorporated as a means. Over time, however, the separation of the material and moral continued to prove fit, while the purposed harmony between them, for lack of evidence, waned. During the nineteenth century, material well-being steadily rose to become an end in itself. The transformation of nature into stuff for us to consume became evidence of our scientific progress and technological prowess—no longer a means but an unassailable manifest destiny.

We began to believe the dream of early scientists that we could control nature to serve our ends. And with so much material progress, we fell into the curious situation of thinking we did not have to worry about living with nature or saving it for future generations. With science assuring a better life for our children than we were experiencing ourselves, moral discourse and historical teachings about responsibility toward both the planet and future generations withered. Going one step further, moral concern for the poor diminished with the argument that the next generation would be better off anyway.

And then stepping over the brink, our respect for the cultures of others, or even simply our tolerance—not that either was ever in great supply—became obstacles to progress. People who resisted material destinies were slaughtered, infected with diseases, pushed off their land, left homeless, and forced into the few familiar nooks and crannies that remained.

In the ideal world of progress, cultural differences would shrink to no more than the French cooking with more butter and the Thai with more spices. The important stuff, the material stuff—calories, protein, and vitamins—would be made available for all through the application of modern scientific knowledge. Progressive Westerners have argued and continue to argue that the unification of cultures would bring peace. Cultural conflicts would fade. A shared cosmology would make it apparent that there are opportunities for all, or at least those who remained, to work together toward further material improvement.

So great was our belief in progress as material advance that it survived the First World War, the Great Depression, and the Second World War. The Cold War was a struggle over whether decentralized capitalist or authoritarian socialist systems would bring material gains most quickly. Meanwhile, material gain lost its zero point, its baseline—any absolute meaning whatsoever. At the same time, the temporal and spatial consequences of the economic transformation associated with material progress made refrigerators and cars, for example, a necessity. So, on the one hand, what was materially essential kept expanding; on the other hand, happiness came increasingly through having things before your friends did, being ahead of the curve in an endless outward spiral of material consumption. Happiness through friendship shriveled as our time became too valuable—that is, more tightly tied to expanding production and consumption. Pursuing material gain has not had many intended effects beyond longer lives, and yet we are all caught in its squirrel cage treadmill, unable to stop running, unable to question the system we have created.

There must be a fuller way of understanding how things unfolded to this situation, a way of understanding that helps us get out of the mess while reinforcing the potential worth of our diverse histories, complex selves, and nature.

Agriculture arose 7,000 to 10,000 years ago when the human population was about 5 million. The population doubled about seven times by the time modern science arose and eight times, rising to 1.6 billion, by the mid-nineteenth century, when modern science and technology really began to affect food production and human health. Population has doubled another four times since then. Progress is central to our story of these latter four doublings, but what of the earlier eight? Or the many more doublings of human activity before the evolution of agriculture? Knowledge, prediction, the deliberate design of technology and economic institutions to promote material gain are little invoked for our history before the nineteenth century, yet humankind made vast, diverse strides, many of them material. And if material advance through science, technology, and economics is not essential to the earlier story, why do we think they are essential to the latter? Or to our future? Could another process be underway throughout our history? And wouldn't it be nice if this process helped us understand and value diversity and human bonds rather than emphasizing more material stuff?

The question of how we really got to where we are first struck me while working with an Amazon planning team for the Brazilian government. In the late 1970s, in Brasilia far from the Amazon, the planners figured they could rationally deduce how to bring economic development to the tropical rain forest. The planning center had a full set of maps with forest and soil qualities identified by remote sensing. We had a vast library of reports by soil scientists, agronomists, and foresters. We could call in experts as needed. Of course, Brazil's rationally planned efforts to colonize the Amazon over the past decade had failed miserably. Few were curious as to why Western development of the Amazon had never gone as planned for nearly five hundred years. Certainly none of the planners wanted to go to the Amazon. The planners felt no need to talk to anyone actually administering faltering projects, let alone farmers and ranchers with dirt on their hands. They had their Ph.D.'s in planning from France and the United States and knew quite well that progress in the Amazon would march to properly orchestrated facts. They were living the modern story of how science and economics drives material progress.

Europe and North America were not planned, scientifically or otherwise. Of course, individual parts were thought out, but there were many parts, and a good number of them, if not the vast majority, failed. On a grand scale, development in Europe and North America was a process of trial and error, a lot of experimentation. The projects that survived were those that turned out to fit together and also to meet unexpected needs. Things coevolved. The arrival of trains made farm machinery and large farms serving distant markets more fit and other things less fit. One can see history this way and still argue that a rational process of design, rather than simply initiating many experiments, would have been superior. Surely, it is better to think than not to think about the future—to think and then act as wisely as we can.

The problem really is with how we think over longer time periods and larger spaces. It is quite rational to set my course walking to the university as if the earth were flat, oblivious of the earth's curvature, how earth turns on its axis and circles about the sun. Sending a rocket around Mars requires an understanding of the bigger picture. While serving on the planning team, I realized that developing the Amazon would be better understood as a process of coevolution. If this is so, then surely acting on the premises that we can predict, design, and control—the modern idea of how science drives material progress—will lead us astray. So let me elaborate on the biological idea of coevolution before taking it to the social.

Plants and animals evolve to fit their niche better, but niches are predominantly defined by other plants and animals. Thus, species largely coevolve with each other. Any change in characteristics of predator or prey selects new characteristics in the other. Coevolution describes how interconnected species evolve together. Biologists have always acknowledged that species define each other's niche. But it is always easier to explain evolution as a process of individual species "improving" through the survival of the fittest to suit an unchanging physical niche better and better. "Better and better" sounds like progress, even if it *is* a miserable niche. One way the Western mind could accept evolution was to conflate it with progress. And in this way, we did.

Something critically important happens when you think of the evolutionary process among species. Evolution can no longer be equated with progress. Everything changes without direction. Just as the best predictor of tomorrow's weather is the weather today, what has been dominant affects what will prove fit next. Yet where the system ultimately goes is driven by mutations and the introduction of new genetic characteristics and new species from beyond the ecosystem, and these inputs are random, sending the coevolving system this way and that.

Until merely a few centuries ago, the world was a patchwork quilt of specific geographic regions within which human cultures and ecosystems were coevolving much like ecological systems. Within each patch, environmental characteristics, technologies, social organization, knowledge, and values put selective pressure on each other so that they fit together. The dominant coevolutionary processes occurred within the patches—the geographic regions associated with specific human cultures. Boundaries, however, were neither distinct nor fixed within this coevolving mosaic. Knowledge, values, social organization, technologies, and species spilled out from the areas within which they initially coevolved to become exotics in other areas. Some of these exotic cultural characteristics proved fit as they arrived and thrived in their new locations, some adapted, and some died out. The process was not entirely random. Human will and power were important. There were tyrants from the beginning who forced "introductions" of social organization. Nevertheless, even imposed regimes eventually had to fit into the coevolutionary process so that the expense of forcing them on others would not be too great. Everything that was newly introduced to an area reset the dynamics of the ever recomposing culture and distorted the shape of the patch. The infinite possible combinations of spillovers and rooting of exotics into different parts of the global quilt kept the spatial pattern of coevolving species, myths, organization, and technology constantly changing and blurry.

The rising dominance of Western science, technology, social organization, and values began to harmonize the global quilt around 500 years ago. During the past 150 years, fossil hydrocarbons provided a new energy source, which—with the advent of steam technologies, chemical industries, and later the internal combustion engines—drove a wedge in the coevolutionary process between people and the environment. This seemed to free people from nature, to allow us to forget the endless interdependencies every good partnership must recognize. Social organization coevolved more and more with fossil fuels, and with material progress now self-evident, all soon sought this advantage. Accepting the belief of progress softened the imperialism of Western culture.

Now the simple material gains and cultural homogenization of the past few centuries are mixed with complex consequences. The hydrogen wedge did not halt coevolution; it just shifted it to more indirect and longer-term processes. Climate change and the loss of biodiversity make it clear that we are still coevolving with nature. The species that will be left for our grandchildren will be those that fit the environments we have created. Meanwhile, the homogenization of cultures proved neither so easy nor so desirable as expected. Unceasing outbreaks of war around a multitude of ethnic differences make it clear that unity has not been won and is not in sight. Material development remains more uneven than before. Gross material inequalities seem to be making it impossible to agree collectively on how to manage the global environment, as well as which new technologies are needed. Environmental management is thwarted nationally as capitalists circle their wagons and protect their ill-begotten share. Globally, the rich nations are doing the same. People who should have been enjoying life in the beam of progress are beginning to accept how it flickers. Could it be like this forever? Won't forever be short if we continue to huddle and search for new solutions under the waning beam of progress?

This way of thinking about change provides flashes of insight toward explaining the multiple dimensions of the human predicament. Things are not marching in converging lines toward a single goal. Rather, our belief in progress has blinded us to the enveloping, unfolding complexity. Staring into an imagined brighter unified future, we cannot see the rich cultural colorations and deep shadows of the present. Believing that the separate efforts of scientists would make a coherent whole has prevented us from addressing how to behave when the pieces refuse to fit. Stressing the material, we have lost sight of the important relationships between us and those to come, as well as the ethical rules that sustain them.

Coevolutionary framing helps us see how fossil hydrocarbons and associated technologies detached us from nature in the short run yet left us with more difficult long-run associations. We left the common experience of farming with its familial and communal ties to take up specialized tasks producing new goods, with our labor and products linked in markets. The transformation occurred very unequally among people, leaving many to coevolve around the materially rich in diverse and perverse ways. Beliefs about how we relate to nature arose within our highly fragmented society, beliefs that do not fit the long-term biogeochemical realities behind climate change or our dependence on biological diversity.

We're provided with a new source of hope when we come to understand how things change while becoming tightly interlocked together. It suggests that there are advantages of being tied to nature differently in different places. Shifting to an evolutionary understanding of change, we see how diversity ensures the process—that diversity itself is a good thing. We need not simply appeal to liberal values of tolerance to protect diversity, for diversity is functionally important to the resilience of humanity in the face of an uncertain future. Coevolutionary thinking also helps us see how difficult it would be to go back to a patchwork of cultures. At the same time, we have options. There is still time. Perhaps we should not rush ahead to a nuclear power economy, or a solar photovoltaic one either, and think about living interdependently with how nature captures the energy of the sun. Perhaps Muslims, Jews, and Christians could see value in their differences and learn to coevolve in adjacent and shared spaces. We should constrain the introduction of bioengineered inputs—observe how a few experiments change our lives and nature before we go this direction in a big way. We could hold better discussions of the global economy if we had an alternative vision in mind. Perhaps indigenous voices are more important than we have thought. By shifting to a framework that stresses relationships, we could see that relationships are more important than more material stuff. We could make space for branching, coevolving possibilities in the future rather than trying to impose single-minded visions of progress.

Skot Jonz, *Buddha on Ice* (detail)

The Abundance of Less: From *A Different Kind of Luxury*

Andy Couturier

Far into the interior of the sharply cleft mountains of Shikoku, if you follow a steep dirt path past terraced rice paddies, rough stone retaining walls, and through the shade of dark green stands of bamboo and cedar, you will eventually come upon the small wooden cottage of Nakamura-san, the artisan. Standing in front of his house, now out in the clear, you can look across the narrow valley and pick out, on the opposite side, a few other tile-roofed farmhouses set among the tapestry of blue-green trees that cover the mountain. Several hundred feet below, a boulder-filled creek cuts along the valley floor.

Nakamura is someone who is almost always at home—except when he's studying in the tiny Sherpa village in the mountains of Nepal where he learns woodblock printmaking from his teacher, a Tibetan Buddhist monk. So when I stop by late one winter's morning eager for another lively conversation with my friend, I find it odd that he's not around. He has no telephone, so one just has to climb the path and see if he's in.

I leave a note in the mailbox and traipse back down the path to where it meets the road. Walking along, I watch the creek spill into deep green and shadowy pools. I then cross a small bridge and head down hill to the local public hot spring. When I enter the bathing area, I am surprised to find the shaven-headed gentleman himself soaping up in the adjacent stall.

"Yes," Nakamura says, "during the winter, I only go out about three times, and that's usually to the hot spring. With the snowfall last night, the pipes to my bathtub are frozen, so I decided to come down here."

As we soak in the mineral-rich water, he invites me back up the mountain for some locally produced tea—"Awa *bancha.*" As we walk back up the road, Nakamura tells me that this special fermented tea is now made in only two remote mountain regions of Japan and, interestingly, rural Burma.

When we arrive back at his house, Nakamura slides back the perfectly maintained paper shoji doors, and we step into a dirt-floored entry room typical of many old farmhouses in Japan. Less typical is the mud-colored Nepali-style cookstove on which Nakamura prepares all his food. We take off our shoes, step up into the living area, and sit down on rough straw mats around the traditional Japanese sunken fire pit in the middle of the floor.

Nakamura builds a fire well. He places several logs in a cross pattern in the ashes, between which he arranges crisp, dry cedar needles that he takes from a woven straw basket. He then puts a few pieces of split bamboo on top of the cedar, and finally some short, one-inch-diameter branches. Each of these items has, over the preceding year, been gathered, cut to length, dried, and stacked neatly by size and type under the eaves in front of his house.

As always, the fire lights on the very first match.

Being with Nakamura inevitably provokes questions: Why collect wood painstakingly, dry it, store it, and then deal with the inconsistencies and smoke of a fire when propane gas or electricity can provide all one's needs for cooking or heating? And why ignore and reject all the achievements of science and technology bought at the cost of great intellectual and physical labor by engineers and laborers in mines and oil fields in order to live a way of life that millions of people were overjoyed to leave behind? The full answers to such questions, as with most things with Nakamura, come slowly, in time.

Nakamura has lived in this house for almost eleven years now, since returning from Nepal at age forty-two. When I first visited him here, I imagined that his apparently austere way of life was based on some moral or ethical philosophy.

"No," Nakamura says, "everything I do is because I completely enjoy doing it this way."

Nakamura places an elegantly shaped cast iron kettle on the hook that hangs over the fire pit and when the water in it starts steaming, takes a small handful of dried, olive-colored leaves from a tin canister and places them in a mottled, russet-colored teapot. He pours the boiling water slowly over the tea leaves in a circular motion with as much care as one might take when transplanting a seedling. He then reaches over to the wooden shelves behind him for a second canister containing another specialty, this time of Nepali origin: hard, twisted biscuit-like snacks made according to a recipe he learned from the Sherpa villagers. He tilts the canister, and five or six slide out onto a small earthenware plate. Each one is coated with unrefined sugar, is delightfully crunchy, and makes an excellent complement to the fragrant yellowish-green tea.

As we warm ourselves by the fire, the smoke billows and gathers around the blackened roof timbers above our heads. We look out between the open sliding doors on the snow-dusted mountains, and he begins to tell me of the travels that eventually brought him to this life that he now leads.

"Ever since I was a child," he says, "I always looked at the hills behind where we lived and wanted to know what was on the other side. Many people are satisfied with where they are and with what they have. I don't judge that at all, but as soon as I was able, maybe seven years old, I got on a bus by myself to see what was in the next town. As I got older I would go farther and farther away. Eventually I had seen a lot of Japan, but I still wanted to know what was across the ocean. By that time I was working in a factory and had a steady income, but I had that same yearning. I wanted to know what was on the other side of the ocean."

Like the minute geometric designs he chisels into wooden printing blocks, Nakamura's way of speaking is both precise and evocative, but not overly serious. His thoughtful face often breaks up in animated laughter when telling his travel stories, and he can easily spend four or five hours with you in conversation, just relaxing on the straw mats and drinking tea.

"The decision to actually leave happened in an instant," he continues. "I looked at my life, and I knew that I didn't want to wake up one day and find myself an old man filled with regret that I hadn't seen the things of this world."

"Of course," he continues, "there are two kinds of regret I could have faced: I knew it was quite possible that I might end up stranded in some foreign country, miserable, without any money and knowing that I had given up my job. But when I compared that possible regret against retiring at sixty-five years old, having known nothing except working in a factory—that was when I knew. I thought about it for a long time, but the decision came in an instant."

A choice made by an individual, a choice to have his life take on a different meaning. In an economic system of discrete labor, in which each person contributes an ever smaller part to an ever vaster manufacturing of "goods," the act of disentangling oneself from the whole might appear revolutionary. But for the person concerned, in that moment, it feels like the only available path.

When Nakamura quit his job, he was twenty-eight. He left Japan and boarded the Trans-Siberian Railroad, and changing trains in Moscow, came down into Europe. He spent the next two years riding a bicycle from northern Europe to southern Italy, then all the way back up and through most of Eastern Europe and then into Greece—a total of 14,000 miles on bike.

"At one point," he tells me, "I ran out of money, and began to look for work. I was in Sweden, and my money was going fast. Every day I went to *fifty* different shops asking for work. And each place just told me 'No.' Every night my stomach hurt from hunger and especially from stress—I was so worried about what I would do. I even started, in the back of my mind, to think about stealing food, just to get something to eat."

"Why didn't you just write to your parents," I interrupt, "and ask them for money?"

"No," he answers simply. "I did not consult my parents before I left, so I wasn't going to ask them for any kind of assistance."

"But," he continues, "just before I completely ran out of money, someone gave me a job as a dishwasher." Then he adds, almost as an afterthought, "I decided at that time that living a life focused on money was *definitely* not for me. Thinking about money all the time just twisted my personality too much."

Most people after such an experience, it occurs to me, would probably decide exactly the opposite. Perhaps it is Nakamura's attempt to detach himself from money that shows how deeply radical he is. Yet it was not until he visited the high mountain villages of the Himalayas that he first encountered people existing almost totally outside the cash economy. In the end, he spent most of the 1980s living in a Sherpa village in a small mountain hut.

"Sometimes when you think back and reflect upon a particular turn your life has taken," Nakamura muses now, looking up from the fire, "—say, if you had not realized as you were walking out of your guesthouse in New Delhi that you had forgotten something, and gone back to get it, you would not have met that man in the lobby who suggested that you talk to that Buddhist priest in Katmandu about traditional crafts, who, when you met him, told you about a woodblock print teacher living in the mountains. And if you hadn't gone to study woodblocks, you wouldn't have learned the attitude toward living that has allowed you to lead the life you now do. It all seemed quite ordinary at the time, but looking back, you see that your life would have been completely different if you had not forgotten something in your room."

Now it's getting on toward evening, and Nakamura asks if I would like to spend the night. Having nothing to do the next day, I happily accept. We then go back down to the dirt-floored entry room with the cookstove, and he selects the items for dinner. I know from my previous visits that I am in for a treat. Nakamura is one of the most accomplished—though understated—cooks that I have ever met. Cooking and eating are not things he does because he has to: they are central to his life.

"I am reading *Walden Pond* now," he says, "and though it's interesting, I was hoping that there'd be more about growing and cooking food. A life in and with nature should be 50 percent about food.

"Also, it's strange: Thoreau lived there for only a little more than two years. It's more like he moved there in order to write the book. But someone needs about ten years, I think, to understand a particular way of living. Thoreau seems like he was more of a tourist."

I laugh. I've never heard such an unorthodox opinion about the great man, who, to most people's thinking, pioneered the idea of an urban person intentionally choosing a rural life. For us, Thoreau discovered an important truth very early on: urbanization and mechanization, under the guise of "progress," have a tendency to create alienation, a feeling that our lives lack fulfillment and satisfaction (which, at the end of the day, is what most people want).

Nakamura now goes into the back room and brings out a book, *Rhythms of a Himalayan Village*. "This book was written about the village where I lived," he says, pointing to a black-and-white photograph of a misty, barren mountainside with several mud and tile houses scattered along the ridgeline, and distant snowy peaks towering in back. "This is the monastery where I studied woodblock carving, and," he turns the page, "this man is my teacher, Tapkhay Gendun." He points to a photo of a monk sitting by a window, his face lined by years of cold-weather living and meager conditions.

Paging through the book, I lose myself in its beautiful photographs juxtaposed with enigmatic quotes from the Sherpa villagers and passages from Tibetan Buddhist scriptures. As I read, I begin to sense how the villagers' way of living—in its utter simplicity as well as in its depth of meaning—could give Nakamura a physical, bodily understanding of how to live a life of reduced desires and reduced consumption when he returned to Japan. Living five days' walk from the nearest roads or electric lines on a windy ridge top covered in stunted grasses and rock, the villagers show both a physical strength and—through their words—a humility, a steadiness, and an unmistakable wisdom.

"It was because I saw the Sherpas' way of living, and had lived it myself," says Nakamura, "that when I came to this valley in Japan and first saw this house, I knew I could live in the same way right here."

Now he switches on one of the three light bulbs in his house and starts to grind together cumin, ginger, salt, garlic, and red pepper in a rough-hewn stone bowl. Then, starting a wood fire in the low clay cookstove, he places a blackened frying pan on top, cuts up some onions into tiny cubes with a well-sharpened knife, and slides them into the oil in the pan. He puts a great deal of intention into each action: adjusting the fire, carefully eyeing (but not measuring) the quantities of each item. It's a pleasure to watch. He then adds some cut potatoes, and some cabbage, and then the spice paste from the heavy stone bowl.

As I watch, I think how funny and wonderful it is to be this far back in the mountains on a cold winter's evening about to eat potato vindaloo when in the closest city (more than two hours away by bus) it is impossible to find Indian food at all.

Though he lives in a remote area, Nakamura is far from uninformed. During dinner, our discussion meanders through topics such as the transition of the Japanese economy from manufacturing to information processing to his theories about representational versus abstract art. He gets a lot of his information from the excellent national radio station, NHK, on which one can often hear audio essays and literary forums of sophistication and grace. When I ask him what he usually reads, he says, "I like this literary and arts quarterly, *Ginka*," and picks up a thick, illustrated magazine sitting beside his table. The date is 1973.

"The tone has not been as elegant in recent years as it was in the '70s. That's why I tend to reread the issues from that era rather than buy any new ones."

I smile to myself. Nakamura is someone who puts quality and depth before some vaguely defined idea of "cutting edge." So many people believe that whatever is newest must be best. But when we are told by traditional peoples that we must follow the ways of the elders, what does that mean for us who live in industrialized nations? How do we take from the past to build a future that we want? What *is* cultural survival?

To feel and know what was understood before—both as individuals and as a culture—we have to make a conscious effort to reimbibe it. We need to read old books again, to sing the ancient songs; we have to reabsorb and reintegrate into ourselves the brushstroke techniques, the coming-of-age stories, and the methods of farming. For human beings are much more fountain than stone, keeping the same rough shape over time as different water flows through.

Because we live in the flow of time, we will necessarily get alienated from what happened last year, and even more so from what happened last century. However, when we apply the blunt instrument of "progress" to the task of making choices about the future, this choice has a profound impact on both our present consciousness and the world we create.

At present, we are saddled with a misunderstanding about progress that disavows the treasures of tradition, attacking them from the dais of the grand narrative of the "march of humanity" toward a luminous future.

When I first met Nakamura, I had an idea of "the traditional life" (which I thought he was leading) as something that does not change. I have come to see, in talking with him, that tradition is really a way of feeling and understanding that an individual human is unable to accomplish in one lifetime. It is the accumulation of all the new and revolutionary ideas pioneered by the generations before, a product of numberless people living with one another, living with the earth. Tradition is not born full grown from someone's head. Tradition is a continuous accretion of insights, understandings, and ways of doing things. To solve the problems of weather and soil, to carve masks from wood, to alleviate suffering and give comfort: our techniques for these are the product of the steady layering of experience and intelligence of thousands of generations. If we want to keep culture alive, we cannot interact with it as if we are museum curators preserving an artifact. Our habit, it seems, is to put the item, technique, or practice in a glass display case, defended against decay with temperature and humidity controls. But if we ossify a tradition by making it "other," making it static, if we label it and date it and put it under a halogen bulb and call it "the past," its very essence is abrogated.

For even in the past, these tools, practices, songs, or techniques were never "nailed down" but existed in flux, part of an entire way of living. When we try to freeze them in place, time and culture still flow on around it. Eventually the objects or ways of doing things feel archaic and irrelevant—especially to the young.

Nakamura, I believe, has found at least part of an answer. When he chooses to cook on a mud stove with firewood, to grind his own wheat to make flour, to read old books and literary journals, he is making the choice to try to recover the original sensibility, to interact vitally with the past. Yet he doesn't just study it; he tries to understand it on a physical, experiential level.

And now, as I sit at Nakamura's table and compare his absolutely simple, almost bare existence to the sophisticated level of his thought, I feel an admiration for his decisions about what to prioritize in this life. For all the time he spends cutting and gathering firewood, growing food, cooking, or just gazing into the fire, it doesn't seem that his intellectual life has suffered in the least. It is as though the mastery that he has achieved as a craftsperson suffuses all the other spheres of his life.

After dinner, we go to the far back room where Nakamura carves his woodblocks. He reaches inside the handsome antique *tansu* chest, pulls out his most recent work, and puts it on a low wooden table. It is a collection of prints that he has mounted on traditional Japanese mulberry paper and then hand-bound, exquisitely, with a dark indigo-blue cloth cover.

It opens accordion style, the way many books once did in Japan. Each set of pages reveals, on one side, a rectangular sample of rough-woven Nepali cloth printed with a red and black geometric design and, on the other, a rendition of that same pattern by Nakamura as a woodblock print on handmade paper. Nakamura's reproductions of these designs tell of his characteristic precision. He uses the same color scheme as in the Nepali cloth, but the flawlessly matching flywheels, rosettes, chevrons, zigzags, and diagonal inscriptions—because of their precise fit—merge and shift such that my eyes start to play tricks on me, as if I were viewing an optical illusion.

This fine book is at once a work of cultural preservation and of interpretive change. In the Nepali cloth samples, the lines of the black and red patterns don't fit exactly with each other, but the cloth nonetheless exhibits a strangely beguiling beauty, one that can be found only in folk items that are made to be used. It seems to speak of the life of mountain peasants and of a way of living that has existed for thousands of years.

Nakamura now reaches into the *tansu* chest and selects several of the dense rectangular wooden blocks that he used to make these prints. Each one has rows and rows of intricate, identical patterns, and each pattern is cut deep into the thick, solid block. A single slip of the chisel, I realize, would ruin the pattern, and he would have to start over from scratch.

I ask him if this repetitive, unforgiving work sometimes gets on his nerves.

"Well, it's true that making something of minute detail takes a lot of focused concentration, and at first you do feel some tension because you are worrying about making mistakes. In my experience, such feelings can continue for a half-hour or more. However, if you keep working, all of a sudden you slip into a timeless space, where the work and you cease to be separate. There's only the work itself. When you come out, you don't know whether several minutes or several hours have passed. I think it's a particular feature of this kind of work.

"A craftsperson's job is half meditation, half creation. It takes creativity to design whatever you are working on, but it takes meditation to do it right. Making things with one's own hands cultivates a certain generosity and openness of the heart. It nourishes that state of mind in the craftsperson, which is intimately connected with an entire way of life."

I am reminded then, with sadness, of the epidemic levels of depression in my own country and wonder whether it might have to do with the aversion we have as a culture to working with our hands. Perhaps, for people in industrialized societies, it isn't that "manual labor" is intrinsically unpleasant, but we get frustrated because our attitude is one of resentment toward something demeaning. Viewed differently, however, such work presents us with an opportunity to know ourselves and the physical world itself better by exploring this essential aspect of being human: our relationship with our hands.

Nakamura shows me a diagram that is used for practice by the beginning student of Tibetan woodcarving. Instead of images or patterns, it's a chart of Tibetan lettering, with arrows indicating the correct angle and pressure that the chisel should take. The Tibetan script is tremendously attractive. It has the flow and sweep of Arabic, but also an angular, blade-like assertiveness. It moves from right to left and drops down like curved icicles from a ledge that runs along the top.

"Learning how to accurately carve all the letters of the Tibetan script," says Nakamura, "gives the learner all the skills with the eyes, fingers, and hands that he needs to be a proficient wood carver."

Nakamura pulls from the cabinet some other books that he has bound by hand and shows me the Japanese method of sewing together the cloth-and-paper covers. I look at each of them and marvel at the care that went into them. Given how much labor they take, I realize that it is possible to make only a few copies of each and that only a few people will ever see them. It seems a lot of effort for very little reward. But in contrast to a book published by machines in a factory, even one selling tens of thousands of copies, the simple potency and beauty of a hand-sewn book gives the reader pleasure of an entirely different order.

One of the books Nakamura has hand-bound is nothing more than a few photocopied pages on how to weave sandals from rice straw. Looking at it, I see how much my way of thinking about "craft" has changed over the period I have known him. Instead of it being a "nice" pursuit with which to fill some bored hours around the house, I understand it now as one of the most fundamental and deeply ancient ways that humans have to meet their needs: baskets for winnowing grain, woven cloth to cover the body, forged and hammered iron tools with which to cultivate the soil, and woodblocks to communicate in words and images. I also have seen, in spending time with Nakamura, that a person, in the process of making something like straw sandals or a handmade book, cultivates humility while doing something that is incontrovertibly meaningful.

And yet, I still wonder, as he starts to explain to me the painstaking process of sewing the cover on a book—how the cloth has to be tucked in at a certain angle under the paper, how the cover should extend just 1 millimeter beyond the stack of pages—whether it's really necessary to be *that* careful.

Nakamura's answer is simple, and clear: "If you make it this way, every time you look at it later, it's an enjoyable experience." It occurs to me then that the energy he puts into making something properly continues to feed his spirit time and again.

Looking at the book on how to make sandals from straw, I see that through his binding these photocopied pages in a cover of black and red Nepali cloth, they have become something of beauty where simply functional would easily have done. I think of my own piles of papers in stacks and in boxes all over my house, and compare them less than favorably to the grace and simplicity of his way of keeping information he deems important.

When I share this thought with him, he replies by walking over to his bookshelf and pulls down a tiny book covered in yellowed kraft paper. "If I were to leave this house and this way of living behind tomorrow, this is one of the three books I would take with me. It's entitled *The Culture of Handicrafts* and was written in the 1920s by Muneyoshi Yanagi. Yanagi asked himself the question: 'What is beautiful?' And the answer, for him, was: 'Everyday things; things that are used in daily life by ordinary people.'"

Yanagi, I learn, was one of the pioneers of the Japanese folk handicrafts movement, which rejected ostentatious ideals of beauty held by many Japanese of the time. He traveled throughout the Japanese islands, and Korea as well, living with peasants and farmers and looking for rough earthenware rice bowls, handmade wooden buckets, bamboo and straw baskets, all of which, to him, expressed a more subtle type of beauty.

Here I find the foundation that underlies Nakamura's approach to life: in Yanagi's words, "The Beauty of Usefulness."

As I look again around the three rooms that make up Nakamura's house, I'm able to take in its beauty in a new way and appreciate the aesthetic of a house seemingly empty of everything except the smoke from the fire. Next to the *irori* firepit, where we have now returned to rekindle the fire for another cup of *bancha,* I look more carefully at the simple fire implements, the several woven wheat-colored kindling baskets, and a blowing tube of bamboo for bringing embers to life. Each object, I realize, he has chosen with deliberation.

In the adjacent cooking area, back in the shadows, stands a heavy ceramic, nutmeg-colored urn almost three feet high in which he stores his water. To the right of the sink, there is an ochre ceramic vessel with a fitted lid containing ash from the fire, which, along with a stiff, natural bristle brush, is all that he uses for washing dishes. At the far end of the darkly stained wood dining table, amber-colored wooden shelves hold glass jars of different sizes containing spices, powders, chili peppers, and several types of flour, all lined up and classified by size. A well-proportioned cast-iron spatula with a long, tapering handle and a blade shaped like a ginkgo leaf hangs on a post in the center of the shelves. Behind the shelves, his cereals and grains are stored in glass jars in large hanging straw baskets suspended from the ceiling.

Nakamura keeps his small collection of books (each of which has been covered in yellow kraft paper and labeled in his distinctive hand) in the back room where he sleeps, closed behind the sliding drawer of a wooden chest instead of having them on display in the main room, where the titles on the spines might detract from the sparse sense of emptiness one feels by the fire.

Perhaps most enigmatic of all the things to look at in Nakamura's house, however, is the texture and coloration of the walls and the sliding paper doors. Though much of the rich patina is simply the result of years of smoke from the fire and the aging of natural materials—wood, paper, mud, and straw—the effect is that of a fine piece of art, a kind of supplication to the senses. Starting from the dun-colored mats on the floor, the mud walls and rice paper shoji panels change hues in a subtle gradation, from a parchment yellow and moving through amber to dark tan and then to a thick caramel brown where they meet the blackened timbers that run along the ceiling. The tannins and oils released from the burning of the cedar and oak firewood have left a speckled pattern on the walls, accumulating gradually like fine siltation on the shore of a lake.

The water is again boiling, and Nakamura takes out more tea. Smelling the fragrance from the teapot, I think back to the previous summer when I came up to the mountains to learn the process of making fermented *bancha* with Nakamura and some of the people in the village. In the heat of July, we picked the large, rough leaves from bushes on the mountain slopes and then transferred them from the straw baskets we wore on our chests into boiling cauldrons of water. After crushing the leaves with a long-handled wooden press, we packed them into large ceramic jars and covered them with stones, so they could cure for a month. Many of the older village residents have been making this tea in exactly this manner since they were small children, when almost everything was, of necessity, done by hand.

In Japan, I have learned, even up through most of the 1950s, most rural people had to make do with little more than what they could grow or make, hardly using cash at all. The mountains themselves, Nakamura tells me, used to be the central resource for people all over Japan, providing them with wild vegetables, medicinal herbs, materials for tools, fuel for their fires, and thatch for their roofs.

But now, with cars and the almost universal use of cash, the mountains are perceived as no more than an inconvenient location for the houses of those people unfortunate enough to have to stay there for one reason or another. Over the years of Nakamura's life, millions of Japanese have fled the mountains and other rural areas to live in the vast cement conurbation that stretches for hundreds of miles from Hiroshima to Tokyo. They exist among a forest of vending machines and petty gambling *pachinko* parlors lit up with bells and sirens and flashing lights. The numberless blemishes of dams, factories, power lines, containerized shipping ports, cemented-in rivers, robotized factories, and fleets of roaring earth moving equipment: all share the soil with the citizens of the world's second largest economy.

And yet, though the values that Nakamura lives by seem to be antithetical to the majority of Japan, he is careful to explain, "It is only because of the shift in values and lifeways in rural areas that I am able to live here at all. Fifty or a hundred years ago, it would have been impossible for an outsider like me to have come here and lived: this house, for one thing, would not have been empty, and it would have been impossible for me to use the community water or the community roads because I had not been part of helping to repair and maintain them for all of the previous years. The way of life I live today here in this house, actually, is possible only because I am a small minority. If everyone today lived this kind of a life, I would have to run out early in the morning to gather wood; otherwise, someone else would get it first, and it would all be gone."

Though Nakamura is, in many ways, one of the most self-reliant people I know, he does use cash. For the $4,000 a year or so that he needs, he leaves his village in October and for about forty or fifty days works at a reconstructed nineteenth-century crafts village baking *mochi,* a traditional sticky white rice cake, over a charcoal fire for urban visitors eager for a taste of old Japan.

For his woodblock prints, however, he will accept no money. "If I did," he says, "it might spoil the enjoyment I get from the process of carving." And losing the pleasure he gets from his way of life, it seems, is the last thing he is willing to do.

"In the summer," he says, "when I spend time making this *bancha* with the older villagers or gathering firewood, it's 'work,' but it's not *labor.* Making *bancha* is something that I do once a year, and afterward, I have enough tea to last me a whole year. Wage labor, like I do at the crafts village, is when I hand over my time for money, and I will do whatever is asked of me, whether I enjoy it or not."

As the fire burns down and we stare into the coals, I think about how little he uses, yet how satisfied he is with his life. Even things that Nakamura has in abundance, such as water or garden space, he uses extremely sparingly. He keeps only a few books. As he says, "I can reread the most important ones every two or three years, and as I grow older and change, I receive different things from the same words." It is as if he is training his spirit by refraining from any form of psychological or material gluttony.

Frugality, it seems, does not mean deprivation, and I speculate that the lack of clutter in his house and in his mind gives a feeling of peacefulness to his presence that his visitors can immediately sense. His attentiveness to material things, I have observed, allows him to connect on a constant level to each moment of his day. The firewood all stacked and drying under the overhanging eaves of his roof; the well-built bamboo trellises in the neatly weeded garden; the wood and tin washboard with which he cleans all his clothes. When I first met Nakamura, I thought to myself, "What a burden it must be to do everything by hand, and with this level of order and care." But when I commented on it to him, he replied in his typical manner of cutting through layers of misperception with a single statement, "If you have *time*, a lot of things are enjoyable. Making this kind of woodblock, or collecting the wood for the fire, making the fire and cooking, doing the gardening, or even cleaning things—it's all enjoyable and satisfying if you give yourself time."

"I think," Nakamura continues now, "that human beings have a tendency to create a visual image in their minds, of what they *think* they can accomplish in a particular period of time—say in a day or a week or a year. But one thing I noticed when I first came here was that there is a gap between that image and the amount I can actually accomplish. I felt ill at ease and irritable. I eventually learned, however, to adjust my imagination, and plans, to what was *actually* possible."

Perhaps this is how Nakamura keeps his presence so calm: by reducing the number of plans he makes so that they fit easily into the time that he has available instead of trying to accelerate his life to accomplish a vast list of projects. Maybe he has come to understand this way of living a satisfied life precisely because he has set a pace slow enough to provide room to observe the processes of his own mind.

I remember one of the first times I met Nakamura. He said (and it puzzled me at the time), "I choose this kind of existence as an experiment, as a way to discover the best way for me to live my life." Now, after many conversations, I understand that he is not living according to a set of abstract ideals. As he says, "Everything I do is because I enjoy doing it this way. I could start cooking on a gas stove, but then I would lose the pleasure I get from gathering and splitting firewood. Or if I didn't grow my own food, I would lose the enjoyment of working the soil."

Outside, the snow has turned to rain, and with its gentle sounds on the roof, Nakamura lays out some old futons for me to sleep on. As my body begins to warm up the bedclothes, the familiar tranquility of the Japanese countryside settles over me again, and with the sounds of the river below and the rain falling on the roof, I drift off to sleep.

In the morning, when I awake, Nakamura is already up and making breakfast. Out the windows, mists are gathering and drifting on the dark green cedar forest across the rainy valley, looking like nothing so much as an old Chinese ink painting.

On the cookstove, Nakamura is making some sticky bread muffins in a set of stacked bamboo steamers over a wok full of boiling water. The crackle of kindling reminds me again of Nakamura's wood-centered life. At the table, he is cutting up dates and other dried fruits to add to our *tsampa,* the Tibetan staple food made from toasted barley flour, dried fruit, a chunk of butter, and (in Nakamura's case)

Awa sour tea, which is then mixed into a paste. It's not what I usually have for breakfast, but it's warm and tasty, and the dried fruit makes it particularly delicious.

I look outside, and it seems as if it might just keep raining all day. I ask Nakamura what he usually does on days like today.

"Sometimes I carve woodblocks or read, but mostly when I have nothing to do, I just stare into the fire," he says.

"Doing nothing all day—it's difficult at first." He laughs. Being busy is a habit, he tells me, and a hard one to break. But perhaps such a life—very little production, very little consumption—might be an important part of the solution to the world ecological crisis.

As we look out on the clouds on the far mountains, I ask Nakamura, "Do you feel that you are living a life of luxury?"

"Luxury? No, not luxury. It's an ordinary life. But I do feel an abundance, a sense of plenty. A hundred years ago, I would not have been able to choose what kind of life to live. I feel very lucky to be living in this age."

Suzanne Stryk, *Evolution* (detail)

From *How Little I Know*

R. Buckminster Fuller

To comprehend the integral of art and science
As an irrepressible, intuitive creative urgency—
As an artist's need to articulate—
Kepes at Massachusetts Institute of Technology
Made a beautiful demonstration.

He took hundreds of 8″ × 10″
Black-and-white photographs
Of modern paintings and mixed them thoroughly
Like shuffled cards
With photographs taken by scientists
Through microscopes or telescopes
Of all manner of natural phenomena
Sound waves, chromosomes and such.
The only way you can classify
Photographs with nothing recognizable in them
Is by your own spontaneous
Pattern classifications.
Group the mealy, the blotchy, the striped,
The swirly, the polka-dotted, and their sub-combinations.
The pattern classified groups
Of photographs were displayed.
The artists' work and the scientists'

Were indistinguishable.
Checking the back-mounted data, it was found
That the artist had frequently conceived
The imagined pattern before
The scientist found it in nature.
Science began to take
A new view of artists.

Loving mothers
Prohibit here and promote there—
Often in ways irrelevant or frustrating
To brain-coordinated genetic evolution,
Often suppressing
A child's profound contribution
Trying to emerge.
We have to look on our society
As we look on the biological world in general
Recognizing, for instance,
The extraordinary contributions
Of the fungi, the manures, the worms, et al.—
In the chemical reprocessing—
And fertility up-grading of the earth.
We must learn to think
Of the functions of the trees' roots
As being of equal importance
To the leave's functions.
We tend to applaud
Only the flower and the fruit
Just as we applaud only the football player
Who makes the touchdown
And not the lineman
Who opened the way.

What society applauds as "creative"
Is often isolated
Out of an extraordinary set

Of co-equal evolutionary events,
Most of which are invisible.
Evolutionary "touchdowns" are unpredictable—
Sometimes centuries apart.
Who knows which child is to make the next breakthrough?
In the next decade society
Is going to be preoccupied with the child
Because through the behavioral sciences
And electrical exploration of the brain
We find that given the right environment
And thoughtful answers to its questions
The child has everything it needs educationally
Right from birth.

We have thought erroneously of education
As the mature wisdom
And over-brimming knowledge of the grownups
Injected by the discipline pump
Into the otherwise "empty" child's head.
Sometimes parents say "don't"
Because they want to protect the child
From getting into trouble.
At other times when they fail to say "no"
The child gets into trouble.
The child, frustrated, stops exploring.
It is possible to design environments
Within which the child will be
Neither frustrated nor hurt
Yet free to self-educate, spontaneously and fully
Without trespassing on others.
I have learned to undertake
Reform of the environment
And not to try to reform man.
If we design the environment properly
It will permit both child and adult to develop safely
And to behave logically.

Order is achieved through—positive and negative—
Magnitude and frequency controlled alteration
Of the successive steering angles.
We move by zigzagging control
From one phase of physical Universe evolution to another.
The rudder concept of social law is most apt.
The late Norbert Wiener chose the word *cybernetics*
Derived from Greek roots of "rudder"
Because Wiener, Shannon and others in communication theory
Were exploring human behaviors
And their brain-controlled "feedback," etc.,
As a basis for the design of computers—
And it became evident
That the human brain only waveringly
Steers man through constant change.

No sharp cleavage is found
Which identified the boundary between life and non-life,
Between the heretofore so-called "animate" and "inanimate."

Viruses,
The smallest organized structures
Exhibiting "life,"
May be classified either
As inanimate or animate—
As crystalline or "cellular" forms.
This is the level also at which
The DNA-RNA genetic code serves as
An angle and frequency designed
Structural pattern integrity.
Such pattern integrities
Are strictly accountable
Only as mathematical principles
Pattern integrities are found
At all levels of structural organization in Universe.

The DNA-RNA is a specialized case
Of the generalized principle of pattern integrity
Found throughout life and non-life.
All pattern integrity design
Is controlled entirely and only by
Angle and frequency modulation.
The biological corpus
Is not strictly "animate" at any point.
Given that the "ordering"
Of the corpus design
Is accomplished through such codings as DNA-RNA
Which are exclusively angle and frequency modulation.

Then we may go on to suggest
That "life," as we customarily define it
Could be effected at a distance.
Precession is the effect
Of one moving system
Upon another moving system.
Precession always produces
Angular changes of the movements
Of the effected bodies and
At angles other than 180 degrees,
That is, the results are never
Continuance in a straight line.
Ergo all bodies of universe
Are effecting the other bodies
In varying degrees
And all the intergravitational effects
Are precessionally angular modulations
And all the interradiation effects
Are frequency modulations.

The gravitational and radiation effects
Could modulate the DNA-RNA

Angle and frequency instructions
At astronomical remoteness—
Life could be "sent on."

Within the order of evolution as usually drawn
Life "occurred" as a series
Of fortuitous probabilities in the primeval sea.
It could have been sent or "radiated" there.
That is, the prime code
Or angle and frequency modulated signal
Could have been transmitted
From a remote stellar location.
It seems more likely
(In view of the continuous rediscovery of humans
As fully organized beings
At ever more remote historical periods)
That the inanimate structural pattern integrity,
Which we call human being,
Was a frequency modulated code message
Beamed at Earth from remote location.
Man as prime organizing
"Principle" construct pattern integrity
Was radiated here from the stars—
Not as primal cell, but as
A fully articulated high order being,
Possibly as the synergetic totality
Of all the gravitation
And radiation effects
Of all the stars
In our galaxy
And from all the adjacent galaxies
With some weak effects
And some strong effects
And from all time.
And pattern itself being weightless,

The life integrities are apparently
Inherently immortal.

You and I
Are essential functions
Of Universe
We are exquisite syntropy.

I'll be seeing you!
Forever.

Contributors

Carolynne Baker worked in Hanoi as an architect for AusAid's Youth Ambassadors for Development Program. She resumed her postgraduate studies in architecture at the University of Melbourne when she returned to Australia in 2003.

Warren Bennett received his B.F.A. from the Rhode Island School of Design. He recently worked for the American Museum of Natural History painting fish for the newly renovated Hall of Ocean Life.

Simmons B. Buntin is the founding editor of the on-line environmental journal *terrain.org*. He has published in *Southern Humanities Review, Sou'wester,* and elsewhere and is a recipient of the Colorado Artists Fellowship for Poetry.

John Canaday's first book of poems, *The Invisible World,* won the 2001 Walt Whitman Award from the Academy of American Poets. His poems have appeared in *Poetry, Raritan, Slate, Paris Review,* and other journals and anthologies. His is also the author of a critical study, *The Nuclear Muse: Literature, Physics, and the First Atomic Bombs.*

Sharon Carter was born in London and obtained her medical degree from Cambridge University. Her poems have been published in *Raven Chronicles, Mediphors,* and *Spindrift,* and on Metro buses. She is a coeditor of *Literary Salt,* an on-line literary journal. Carter received a Hedgebrook residency in 2001.

Andy Couturier is a writing teacher and essayist. His work has appeared in the *North American Review, Creative NonFiction, Adbusters, invAsian Journal,* and elsewhere. "Living the Abundance of Less" is one of the chapters for his forthcoming book, *A Different Kind of Luxury,* and is based on a series of articles in the *Japan Times.*

Ellen Dissanayake is the author of *What Is Art For?* (1988), *Homo Aestheticus* (1992), and *Art and Intimacy: How The Arts Began* (2000), all of which are concerned with the place of the arts in human evolution and human experience. She has lived and worked in Sri Lanka, Nigeria, and Papua New Guinea and is currently Visiting Scholar at the Center for the Humanities, University of Washington, Seattle.

R. Buckminster Fuller (1895–1983) gained renown as inventor of the Geodesic dome—the only structure that gets stronger as it gets larger. As a philosopher and systems thinker, he coined the term *Spaceship Earth* and organized the World Game. Bucky was World Fellow in Residence at the University of Pennsylvania and received the Medal of Freedom from President Ronald Reagan in 1981.

David C. Geary is the Middlebush Professor of Psychological Sciences at the University of Missouri. He is the author or coauthor of more than 100 scientific publications in a variety of disciplines, including his most recent book, *Male, Female: The Evolution of Human Sex Differences.*

Peter Goin is the author of four books, including *Nuclear Landscapes* (1991) and *Humanature* (1996). His photographs have been exhibited in more than fifty museums nationally and internationally, and he is the recipient of two National Endowment for the Arts fellowships.

Kristjana Gunnars is professor of creative writing at the University of Alberta, Canada. Her most recent book of poems is *The Silence of the Country,* and her latest prose work is *Night Train to Nykøbing.* She has received the Stephan G. Stephansson Award for poetry and the McNally Robinson Award for fiction, and she was short-listed for the Governor General's Award of Canada.

Craig Holdrege is an educator and biologist. He was a high school life science teacher for twenty-one years and is now the director of the Nature Institute in rural New York. He is the author of the book *Genetics and the Manipulation of Life: The Forgotten Factor of Context.*

Aaron Holliday became so obsessed with drawing at the age of thirteen that his mental health was affected and he was hospitalized for three years at the Los

Angeles General Hospital. Holliday still lives in a rather sheltered environment, where he can be alone if he desires. He continues to spend much of his time drawing or painting. Holliday's work has been displayed at numerous art shows and conventions around the country.

Valerie Hurley's fiction and essays have appeared in *New Letters, The Iowa Review, Boston Review,* and other literary journals. She was the recipient of the DeWitt Wallace/Reader's Digest Fellowship at the MacDowell Colony in 2002. Her first novel, *St. Ursula's Girls Against the Atomic Bomb,* was released in the fall of 2003.

Dale Jamieson is professor of environmental studies and philosophy at New York University. His most recent book is *Morality's Progress: Essays on Humans, Other Animals, and the Rest of Nature.* Jamieson is also the editor or coeditor of seven books, including *Singer and his Critics* (1999), named by *Choice* as one of the outstanding academic books of 1999.

Judy Johnson-Williams is an artist who started out as a chemist. She recycles large pieces of cardboard into powerful meditative statements. Her work is collected nationwide.

Skot Jonz's favorite achievement is living the childhood dream of working as an archeologist, which began while he was still in high school. As a field lab director in charge of photographing excavations and artifacts, Jonz developed an interest in photography and has since had annual exhibits of his photos included in collections internationally. He is currently publishing twenty-five years of research on the synchronistic occurrence of numbers and the evolution of consciousness.

Joan Maloof is a botanist whose work crosses the boundaries of science and art. She is the coordinator of Environmental Studies at Salisbury University in Maryland.

Richard B. Norgaard is professor of energy and resources at the University of California at Berkeley. He studies the contradictions between neoclassical economic theory and systemic understandings of nature and society. He is the author of *Development Betrayed: The End of Progress and a Coevolutionary Revisioning of the Future.*

Kathleen Creed Page is associate professor of physiological biochemistry at Bucknell University. Her research focuses on how prenatal stress affects reproduction.

Ricardo Pau-Llosa's fifth collection of poetry is *The Mastery Impulse.* An interview with him appears in the summer 2003 issue of *Manoa.*

David Petersen is the author or editor of a dozen books, including *Heartsblood: Hunting, Spirituality, and Wildness in America* and *Confessions of a Barbarian: Selections from the Journals of Edward Abbey.* "January" is adapted from *Confessions of a Contrarian: In Search of an American Life,* a memoir in progress.

Richard Robinson is an award-winning photographer based in Virginia. His work has appeared in *Smithsonian, National Geographic Traveler,* and the *Washington Post Magazine.* He won the Lowell Thomas Gold award for Travel Photography for his piece "Old Man and the Keys" in *Spirit Magazine* (Southwest Airlines).

Theodore Roszak's most recent books include *The Voice of the Earth, The Memoirs of Elizabeth Frankenstein, The Gendered Atom: Reflections on the Sexual Psychology of Modern Science,* and *Longevity Revolution: As Boomers Become Elders.* His novel *The Devil and Daniel Silverman* was published in winter 2003.

Michael Ruse is the Lucyle T. Werkmeister Professor of Philosophy at Florida State University. The author of many books on and around Darwinism, his latest is *Darwin and Design: Does Evolution Have a Purpose?* This is the final volume of a trilogy in which Ruse looks at the interactions between science and the culture within which it is formed.

Dorion Sagan is a well-known essayist, reviewer, and poet who has contributed to the *New York Times,* the *New York Times Book Review, Wired,* and *Black Bo.* His coauthored books include *Up from Dragons: The Evolution of Intelligence* and *Acquiring Genomes: A Theory of the Origins of Species.*

Eva Salzman's books are *The English Earthquake, Bargain with the Watchman,* and *One Two II,* illustrated by Van Howell. A short opera, *Cassandra,* written with her father, composer Eric Salzman, was performed at the New Opera Conference 11 in Vienna in 2002. Her fiction has been published in *New Writing 10* and *11.*

Cherie Sampson is an environmental performance artist who interiorizes her body in her sculptures in slow ritualistic movement that alludes to the gradual passage of time and change in settings from remote arctic mires to traditional art spaces in the United States and abroad. She teaches digital and performance art at Maharishi University of Management in Fairfield, Iowa.

Andrew Schelling teaches poetry, Sanskrit, and wilderness writing at Naropa University. A translator, ecology activist, and essay writer, he has traveled extensively in India and the Himalayas. His recent books include *Tea Shack Interior: New and Selected Poetry* and a collection of essays, *Wild Form, Savage Grammar.*

Floyd Skloot's most recent book is a collection of essays about living with brain damage, *In the Shadow of Memory* (2003). His fourth book of poems, *The End of Dreams,* comes out in the spring of 2005. Recently, his fiction has appeared in *North American Review, Witness, Tikkun,* and *Virginia Quarterly Review.*

Suzanne Stryk has exhibited her paintings throughout the United States, most recently at Duke University, the Fernbank Museum of Natural History (Atlanta), and Grover/Thurston Gallery (Seattle). She has been the recipient of a fellowship from the Virginia Center for the Creative Arts and the Anselm Atkins Award for the artist inspired by the natural world.

Stephen Miles Uzzo was with the original launch team for MTV and worked for the New York Institute of Technology Computer Graphics Lab and Video Center for a number of years. He currently works for the New York Hall of Science in developing exhibitions and communications networks.

Leslie Van Gelder is the founder of Road Scholars, an experiential travel–based education program for college students. She is currently the co-director of the Oxford Institute for Science and Spirit and is editor of *Green Letters,* the journal of the U.K. branch of the Association for the Study of Literature and the Environment.

Kevin Warwick is professor of cybernetics at the University of Reading, England. His research focuses on intelligent roots. His 1998 book *In the Mind of the Machine* considered how robots may be more intelligent than humans in the future. His 2002 autobiography, *I, Cyborg,* revealed how he became part machine after his nervous system was linked to a computer.

Jessica H. Whiteside, a doctoral candidate at Columbia University, is currently investigating Mesozoic fish systematics and evolution across the Triassic/Jurassic boundary. With Dorion Sagan, Jessica recently coauthored *Gradient Reduction Theory: Thermodynamics and the Purpose of Life.*

Lucila Wroblewski has shown her photographs in Rio de Janeiro, Porto Alegre, Telaviv, and São Paulo. In 1994, she won the Stimulus Prize for her project *Silent Landscape.*

Monika Wuhrer has exhibited her most recent project, *Visceral Circle,* throughout the world, most recently at the Front Room Gallery in Williamsburg, New York, and in Eukabeuk, Thailand.

David Rothenberg is founding editor of Terra Nova and author of *Always the Mountains, Sudden Music,* and *Hand's End.* He is associate professor of philosophy at the New Jersey Institute of Technology.

Wandee Pryor is managing editor of *Terra Nova* and author of the plays *And God Plays Dice, Rats Live on No Evil Star,* and *And Now You Look Like This.*

Sources

John Canaday, "New England Ghazal" reprinted from *The Invisible World,* published by Louisiana State University Press, 2002. Reprinted by permission of the author.

R. Buckminster Fuller, "How Little I Know." Courtesy, The Estate of R. Buckminster Fuller.

Bruce S. Grant, *Biston betularia cognataria,* (The Peppered Moth). This represents an adapted form of an earlier published image in *Journal of Heredity,* Vol. 93, No. 2. Reprinted by permission of Oxford University Press.

Craig Holdrege, revised version of "Science as Process or Dogma? The Case of the Peppered Moth" in *Elemente der Naturwissenschaft* (Vol. 70 (1): pp. 39–51, 1999); reprinted with permission.

Valerie Hurley, adapted version of "Riders on the Earth" in *Missouri Review,* Vol 14, No. 1, 1991.

Michael Ruse, "Is Evolution a Social Construction?," reprinted from *Endeavour,* Vol. 22, Issue 4: 140–142, 1998, with permission from Elsevier.

Andrew Schelling, "Tyger Tyger," appears in *Tea Shack Interior: New and Selected Poetry* copyright © by Andrew Schelling. Reprinted by permission of Talisman House, Publishers.

Floyd Skloot, "The Wings of the Wind" originally appeared in *Virginia Quarterly Review,* Vol. 73, No. 2, Spring 1997.

Kevin Warwick, "Intelligent Robots or Cyborgs" was compiled with the help of excerpts from the following books by Kevin Warwick: *In the Mind of the Machine,* Arrow, 1998; *QI: The Quest for Intelligence,* Piatkus, 2001; and *I, Cyborg,* Century, 2002.